Organic Principles and Practices Handbook Series
A Project of the Northeast Organic Farming Association

Humane and Healthy Poultry Production

A Manual for Organic Growers

Revised

KARMA GLOS

Illustrated by Jocelyn Langer

CHELSEA GREEN PUBLISHING
WHITE RIVER JUNCTION, VERMONT

Editorial Coordinator: Makenna Goodman
Project Manager: Bill Bokermann
Copy Editor: Cannon Labrie
Proofreader: Helen Walden
Indexer: Peggy Holloway
Designer: Peter Holm, Sterling Hill Productions

Printed in the United States of America
First Chelsea Green printing March, 2011
10 9 8 7 6 5 4 3 2 1 11 12 13 14

Chelsea Green Publishing is committed to preserving ancient forests and natural resources. We elected to print this title on 30-percent postconsumer recycled paper, processed chlorine-free. As a result, for this printing, we have saved:

5 Trees (40' tall and 6-8" diameter)
2 Million BTUs of Total Energy
466 Pounds of Greenhouse Gases
2,243 Gallons of Wastewater
136 Pounds of Solid Waste

Chelsea Green Publishing made this paper choice because we and our printer, Thomson-Shore, Inc., are members of the Green Press Initiative, a nonprofit program dedicated to supporting authors, publishers, and suppliers in their efforts to reduce their use of fiber obtained from endangered forests. For more information, visit: www.greenpressinitiative.org.

Environmental impact estimates were made using the Environmental Defense Paper Calculator. For more information visit: www.papercalculator.org.

Our Commitment to Green Publishing
Chelsea Green sees publishing as a tool for cultural change and ecological stewardship. We strive to align our book manufacturing practices with our editorial mission and to reduce the impact of our business enterprise in the environment. We print our books and catalogs on chlorine-free recycled paper, using vegetable-based inks whenever possible. This book may cost slightly more because we use recycled paper, and we hope you'll agree that it's worth it. Chelsea Green is a member of the Green Press Initiative (www.greenpressinitiative.org), a nonprofit coalition of publishers, manufacturers, and authors working to protect the world's endangered forests and conserve natural resources. *Humane and Healthy Poultry Production* was printed on Joy White, a 30-percent postconsumer recycled paper supplied by Thomson-Shore.

Library of Congress Cataloging-in-Publication Data
Glos, Karma.
Humane and healthy poultry production : a manual for organic growers / Karma Glos ; illustrated by Jocelyn Langer.
p. cm. -- (Organic principles and practices handbook series)
"A Project of the Northeast Organic Farming Association."
Originally published: Barre, MA : Northeast Organic Farming Association Interstate Council, 2004.
ISBN 978-1-60358-357-2
1. Poultry. 2. Organic farming. I. Langer, Jocelyn. II. Title. III. Series: Organic principles and practices handbook series.

SF487.G546 2011
636.5--dc22

2010051223

Chelsea Green Publishing Company
Post Office Box 428
White River Junction, VT 05001
(802) 295-6300
www.chelseagreen.com

Best Practices for Farmers and Gardeners

The NOFA handbook series is designed to give a comprehensive view of key farming practices from the organic perspective. The content is geared to serious farmers, gardeners, and homesteaders and those looking to make the transition to organic practices.

Many readers may have arrived at their own best methods to suit their situations of place and pocketbook. These handbooks may help practitioners review and reconsider their concepts and practices in light of holistic biological realities, classic works, and recent research.

Organic agriculture has deep roots and a complex paradigm that stands in bold contrast to the industrialized conventional agriculture that is dominant today. It's critical that organic farming get a fair hearing in the public arena—and that farmers have access not only to the real dirt on organic methods and practices but also to the concepts behind them.

About This Series

The Northeast Organic Farming Association (NOFA) is one of the oldest organic agriculture organizations in the country, dedicated to organic food production and a safer, healthier environment. NOFA has independent chapters in Connecticut, Massachusetts, New Hampshire, New Jersey, New York, Rhode Island, and Vermont.

This handbook series began with a gift to NOFA/Mass and continues under the NOFA Interstate Council with support from NOFA/Mass and a generous grant from Sustainable Agriculture Research and Education (SARE). The project has utilized the expertise of NOFA members and other organic farmers and educators in the Northeast as writers and reviewers. Help also came from the Pennsylvania Association for Sustainable Agriculture and from the Maine Organic Farmers and Gardeners Association.

Jocelyn Langer illustrated the series, and Jonathan von Ranson edited it and coordinated the project. The Manuals Project Committee included Bill Duesing, Steve Gilman, Elizabeth Henderson, Julie Rawson, and Jonathan von Ranson. The committee thanks SARE and the wonderful farmers and educators whose willing commitment it represents.

Organic agriculture is a holistic system of production designed to optimize the productivity and fitness of diverse communities within the agroecosystem, including soil organisms, plants, livestock and people.

—Canadian National Standards
for Organic Agriculture

Organic production is a system that is managed . . . to respond to site-specific conditions by integrating cultural, biological and mechanical practices that foster cycling of resources, promote ecological balance and conserve biodiversity.

—USDA, National Organic Program

CONTENTS

Introduction

I have been a poultry lover my whole life, beginning with a little Barred Rock chick I purchased at the local pet store for Easter. I was only nine, but that ball of fluff started a lifelong interest in poultry. George (though it later turned out to be a hen) slept on my pillow. She was my constant companion and we spent hours wandering the garden looking for worms under rocks. When I had the opportunity to move out to the country to live with my dad, George went too. Dad had an entire flock of chickens: Black Australorps, Rhode Island Reds, and Buff Orpingtons, but George continued to hang out with me. We took daily walks into the hills to explore, and she commonly rode on my shoulder or forded irrigation ditches in my upheld arms. She lived a long life, producing eggs and coming when called.

I have always managed to have contact with chickens and other livestock. When my husband and I established Kingbird Farm in 1996 it was a given that I would keep poultry. Not just chickens—poultry of all kinds, running about the place and troubling my husband's vegetables. We began with a small egg flock, which was quickly followed by our first batch of broilers à la Joel Salatin. This snowballed into layers, turkeys, ducks, and geese. We are still a very small poultry operation by conventional standards, and poultry is only a part of our farm's diversity, but it is a vibrant and viable part of a complex dance on our farm in which plants, animals, wildlife, weather, soil, structures, water, farmers, and chance have their own role in the performance. Our success at creatively and profitably incorporating poultry into this dance has developed from our keen observation of the birds, as we look out for their individual needs and the good of the flock.

one

Organic Poultry Basics

The goal of organic farmers is to develop productive farms that are sustainable and harmonious with the environment. They recognize that there's a relationship between the health and vitality of their animals and the way the animals are kept. They are keenly aware of the links between the health of the land and crops and that of the animals. There's also a "humane" element based on a fundamental respect for the animals, including their behavioral and psychological needs.

Organic poultry production systems can be sophisticated, technically advanced commercial operations or simple, small-farm add-ons integrated into a more complex agroecosystem. No matter the size or scale, the basic tenets of traditional organic agriculture apply.

Creating Humane, Healthful Conditions

Recognizing and respecting the birds' physical, social, nutritional, and psychological needs is a radical departure from convention in this country. Over the years poultry raising has been forced into more and more intensive and unnatural systems geared toward satisfying only the birds' most basic physical and nutritional requirements on behalf of low price and high profit margin. Organic agriculture seeks to return poultry farming to a land-based, humane method that fosters a healthy bird producing a nutritious food. Small- and large-scale organic poultry growers throughout the Northeast and the rest of the United States are choosing to move

Rhode Island Red laying breed.

above and beyond the conventional production model and consciously consider all aspects of poultry husbandry and land care.

A checklist of best practices in organic husbandry in the UK, called "the Five Freedoms," is suited to the ideals of Northeast organic poultry farming.[1]

1. **Freedom from hunger and thirst;** by ready access to fresh water and a diet to maintain full health and vigor.
2. **Freedom from discomfort;** by providing an appropriate environment, including shelter and a comfortable resting area.
3. **Freedom from pain, injury, or disease;** by prevention and rapid diagnosis and treatment.
4. **Freedom to express natural behavior;** by providing space, sufficient facilities, and the company of the animal's own kind.
5. **Freedom from fear and distress;** by ensuring conditions and treatment to avoid mental suffering.

Poultry's Place on the Integrated Farm

Small-scale organic poultry production is complementary with many types of farm production including vegetables, livestock, and orchards. Poultry integrate well with various other farm animals. They contribute by making use of marginal land that cannot be used to grow crops or raise pasture animals and by providing additional income while improving the agroecosystem of the farm. Beyond the obvious benefits of meat and eggs, some farmers believe that raising poultry on their farm aids in the control of parasites in ruminants as well as fly and tick control, though there has been little, if any, scientific research conducted to verify such benefits. Poultry may also play a key role in nutrient cycling. Nutrients are returned to the soil via manure and composted litter. Poultry can be moved around the farm to where nutrients are needed. The farmer's job is to determine how and where poultry best benefit the overall organic farm.

The Intersection of Size, Sustainability, and Profitability

According to *Poultry* magazine, a commercial poultry processors' resource, "High production costs remain an obstacle to moving organic production to the mainstream. Complying with organic requirements is generally not compatible with the practices and routines of conventional broiler production which is achieved through high volume and efficiency."[2] This view is partially correct because many organic poultry producers follow older, traditional methods of animal husbandry. However, due to increased market demand, organic production has also been innovating, and there has been growth in productivity. There are certified organic poultry farms of all sizes in the Northeast from small flocks (fewer than 1,000) on diverse farms to relatively large commercial flocks in barns (107,000 hens at Pete and Gerry's Eggs in Monroe, New Hampshire—perhaps others are bigger) to large packers and processors who contract out to a multitude of small growers. Systems vary somewhat according to size of operation, but all start with the basic principles that organic poultry means free-roaming birds fed organically and raised without conventional medications and physical alterations, such as wing and toe clipping and debeaking. Farms interpret these principles differently, but organic producers, regardless of size, ideally see their farms as living entities with all organisms benefiting each other—not as mono-species factories benefiting only the farmer.

two

Establishing the Facilities

Organic poultry farms utilize a vast array of innovative systems, from completely free-range backyard flocks to vast aviary-style barns that house thousands of birds. When establishing or adapting poultry facilities, it is important to consider what the overall poultry management plan will be. Consider the following elements regardless of what type of bird you may grow:

- What will be your production style? Free-run? Pastured? Multi-species? Movable housing?
- How will you distribute fertility (manure)? On range? Composted?
- What flexibility do you want? Seasonal? Integrated with other production?
- How do you want to use your labor? Will the system be labor intensive or capital intensive?
- What kind of environment do you want to provide the birds? Human friendly, bird friendly, or a compromise?
- What other locational and site-specific issues should you consider? Drainage? Predation? Land (soil) base?
- How will feed be purchased and stored? What rodent and insect control measures do you need?
- How will you collect and pack eggs? Transport and slaughter?
- What laws affect your operation, birds, labels?
- How will you sell your product?

These questions are particularly important for those who wish to establish a sustainable, profitable enterprise while adhering to organic principles. An excellent resource for reviewing your options is ATTRA's publication "Sustainable Poultry: Production Overview" by Anne Fanatico. This article covers alternative poultry production systems and provides important details for how to make these systems work. All of the systems covered can be used for organic production, and many are already in use by Northeast farmers. In order not to repeat the information about housing and equipment covered by Fanatico's publication, I will only cover the basic options here and what might be specific issues for organic growers.

Facilities and Stocking Densities

Given the present provisions of the National Organic Program (NOP) rules, the following examples of production methods can currently be considered when designing organic poultry facilities:

Free-run options[1, 3]

- Aviary-style barn with outdoor access "porch" or straw-bedded yard.
- Deep-bedded hoop house with rotational access to bedded yards.
- Permaculture-style house with rotating access to four yards.

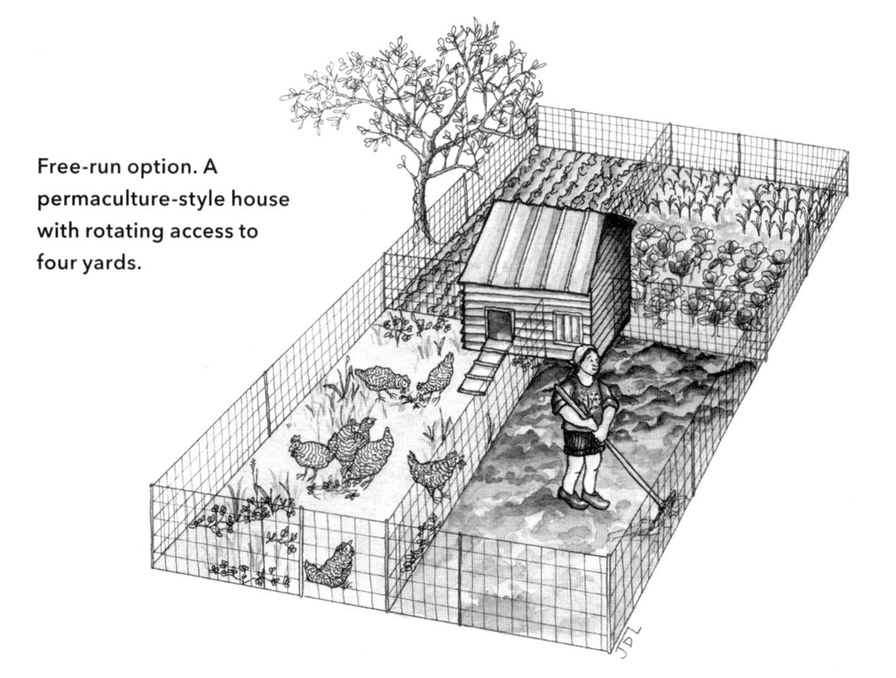

Free-run option. A permaculture-style house with rotating access to four yards.

Free-range options

- Aviary-style barns or hoop houses with "range" access for all of the birds at one time.
- Portable skids, hoop houses, or other structures that are moved around on fallow land, crop residue, or through orchards.

Pastured options

- Stationary barn, hoop house, or other structure with rotating access to pasture managed for the nutritional needs of poultry.
- Skid houses, portable hoop houses, or other shelters, periodically moved across managed pasture within a fenced area (typically electric nets).
- Portable "pasture pens" moved daily across managed pasture.
- Colony systems on managed pasture that include floorless roost houses, feed houses, and bedded nest houses.

Most of these systems are adaptable, with minor adjustments, to all species of poultry. No matter what system is chosen, certain space requirements apply for birds to prosper. Though current U.S. organic standards are sometimes being interpreted to require a minimum of 1.5 sq. ft. of floor space per "bird" (not specific about species) within housing, there are many other space issues important to the physical and overall health of the birds. Birds of all species need room to eat, exercise, flap, socialize, dust, escape bullies, forage, scratch, roost, etc. The following are recommendations based on several authorities and should be adjusted for each production system.

As can be seen in the standards outlined in tables 1, 2, and 3, other countries have much more specific and stringent space requirements for the certification of organic systems. Producers in the United States have been forced to interpret the vague space requirements in the NOP, sometimes with very little guidance or support from certifying agencies. Semi-intensive producers using barns for thousands of birds typically report they feel the space requirements too large and the outdoor access unfeasible or even detrimental. Smaller free-range or pasture-based producers often consider them merely a minimum and provide not only outdoor access, but high-quality managed forage on range. Some producers view

Table 1. Suggested Indoor Stocking Densities Based on Weight	
Fixed housing	
Broilers & ducks	21 kg/sq. m (4.28 lbs./sq. ft.)
Turkeys & geese	21kg/ sq. m (4.28 lbs./sq. ft.)
Mobile housing	
Broilers & ducks	30 kg/sq. m (6.11 lbs./sq. ft.)
Source: "Organic Poultry Production," Soil Association Questions and Answers, Soil Association Producer Services (Bristol, UK, 2003), 10.	

Table 2. Maximum Outdoor Stocking Rates–Birds per Acre					
Layers	**Broilers**	**Turkeys**	**Ducks**	**Geese**	**Guineas**
2,470	6,175	1,976	4,940	1,482	6,175
Source: "Organic Poultry Production," Soil Association Questions and Answers, Soil Association Producer Services (Bristol, UK, 2003), 4.					

Table 3. Space Requirements for Housing	
Broiler chickens	
Chicks to 4 wks.	1 sq. ft./bird
Over 4 wks.	0.36 sq. ft./lb (2.78 lbs./sq. ft.)
Layer chickens	
Chicks to 16 wks.	0.35 sq. ft./bird
Over 16 wks.	2.5 sq. ft./bird, 1,000 birds/acre
Ducks and small game birds	
Floor space	0.36 sq. ft/lb. (2.78 lbs./sq. ft.)
Combined floor/outdoor	40 sq. ft./bird over 10 wks.
Turkeys, geese, and large game birds	
Floor space Apr.-Dec.	4 sq. ft./bird
December-April	0.36 sq. ft./lb.
Outdoor space	80 sq. ft./bird over 10 wks.
Source: "The Organic Production Standards of British Columbia," sections 9.3.2–9.3.4.	

the space and outdoor access requirements as impediments to production or even a danger, exposing their birds to potential pathogens or inclement weather and exposing the land to a bird density it cannot support. Others, however, see them as a bare minimum and a jumping-off place to develop new (or even rediscovering old) and better ranging systems while preserving the health of the birds and the quality of the environment.

A survey I conducted in 2002 (SARE Project FNE02-412) of over forty organic poultry producers in the Northeast found small farms using a variety of the above systems to raise birds within their own diversified farming approach.[2] Every respondent with flocks between 25 and 500 birds provided them with outdoor access. Most provided managed, rotated pasture while a few of the small flocks were given only yards. These producers typically used separate summer housing on pasture and winter housing in barns or greenhouses with access to a yard in all but the most inclement weather. Birds were given ample floor space in both types of housing with an average of 4.72 sq. ft./bird in summer housing and 4.62 sq. ft./bird in winter housing. As evidenced by this survey, smaller commercial organic flocks that are only part of the farm's diversity tend to receive more space and range options. The smaller flock size makes this more feasible, but it also signals the commitment of these farmers to free-range, pasture-based poultry production.

Housing and Equipment Management

Choosing and Using Litter

Regardless of the type of housing, all birds require similar equipment and facilities. Important, particularly in winter housing, is litter management. Most organic producers who answered the survey, whether large or small, typically use a deep-litter system. The use of deep litter in housing provides for the absorption of manure, but also gives a variable substrate (uneven footing) for the physical and psychological health of the birds. Proper litter type and management can provide birds with the following benefits:

- A variable substrate for proper leg development and foot health.
- A friable surface for scratching, foraging, and dusting.
- An absorbent bedding to control manure and other moisture.
- A properly composting pack providing warmth while breaking down manure, destroying harmful bacteria, and controlling ammonia.[3]

Consider these points when managing litter:

- Choose one that is available locally and is economically feasible.
- Options include sawdust, wood chips, wood shavings, hay chaff, soybean stubble, corn fodder, ground cobs, leaves, rice hulls, peanut hulls, chopped pine straw, or chopped straw.
- Hardwood shavings and sawdust often have a high moisture content and are susceptible to dangerous mold growth (this can also happen to improperly stored dry shavings).
- Hardwood chips can cause splinters in birds' feet.
- Avoid cherry and walnut products as the tannins may cause poisoning.
- If hay or straw aren't chopped, they may mat with manure and cause capping.
- Litter that has gotten wet in storage is useless for moisture absorption.
- If litter is eaten by certified organic birds (hay, straw, hulls, etc.) it must be from a certified organic source and comply with the feed requirements of section 205.237 of the final rule (see appendix 7).

A well-managed composting litter pack also provides the farmer with a nitrogen-rich resource, not a pollution liability. In order for this to be true, however, it must be managed with the intent to create a nutrient resource while enhancing the birds' environment. Proper carbon/nitrogen ratios, mixing, turning, and heating are all important aspects to using a composting litter system. The author can recommend the following system for a composting litter:

1. Start with 3–6 in. of dry material on a dirt, cement, or wood floor.
2. Add litter or water as needed to maintain a litter moisture content of 20–25 percent. (To assess the litter condition and moisture content, pick up a handful and squeeze it tightly, then open your hand. If the condition is correct it should feel dry and friable in the hand, and not ball up.)

3. To stimulate bio-processing, encourage birds to scratch up and aerate litter by throwing whole grains into it.
4. Composting litter ideally has a carbon-to-nitrogen (C/N) ratio of 25–35 to 1. It requires mixing ample amounts of bedding—the carbon source—with the raw manure.[4]
5. Periodically rototill or fork by hand (or use a pig) to prevent capping or crusting.
6. When properly composting, the litter stabilizes floor temperature at 70°F.
7. Periodically remove the compost pack. This may only be necessary every two to three years.
8. Further composting outside the poultry house may further reduce pathogens and improve the carbon-to-nitrogen ratio.

A farmer who maintains a certified organic operation, or follows the National Rule without actual certification, must follow further processing and spreading requirements. According to the 2003 recommendations of the National Organic Standards Board (NOSB) Compost Task Force, composted litter would only be considered compost if (*a*) it is made only from allowed feedstock materials; (*b*) the compost undergoes an increase in temperature to at least 131°F (55°C) and remains at that temperature for a minimum of three days; and (*c*) the compost pile is mixed or managed to ensure that all of the feedstock heats to the minimum temperature.[5] If these parameters are not met, the litter is not considered compost and must be treated like raw manure. Currently raw animal manure may only be (*a*) applied to land used for a crop not intended for human consumption; (*b*) incorporated into the soil not less than 120 days prior to harvest of a product whose edible portion has direct soil contact; or (*c*) incorporated into the soil not less than 90 days prior to harvest of a product whose edible portion does not have contact with the soil.[6]

THE AUTHOR'S EXPERIENCE: Winter Housing Challenge[7]

Sizing Seasonal Facilities

One of the more difficult aspects of our certified organic laying flock is winter housing. During the growing season, while the hens are out on pasture, the housing facilities in the field (15 × 48 ft. hoop house) are potentially adequate for 500 birds. Since the birds have constant access to a ¼-acre paddock, rotated weekly, they only use the housing for roosting and laying eggs. Giving them a little less than 1.5 sq. ft. per bird of floor space poses no problems for litter-pack management, bird social order, or organic certification.

In the winter we move the birds to a 12 × 48 ft. polycarbonate greenhouse attached to the barn. Although we allow them outdoor access at all times, there is snow most of the winter, and they do not choose to go outdoors. This effectively means our hens are confined to 576 sq. ft. of deep litter pack for approximately five months. In a confinement situation the allowed square footage of floor space per bird must be increased to let them work out and maintain their social order, do their dusting, eating, roosting, foraging, etc., and to allow the deep litter pack to properly absorb and compost the manure.

With too much space, the birds do not keep each other warm efficiently. Keeping all this in mind we have found that, confined to our winter housing with the deep litter pack, the birds need approximately 2 sq. ft. per bird. This limits our winter population to 300 mature layers. Although our summer housing could accommodate 500, we currently maintain a flock of only 300 hens; but fall culling of the summer flock would also make sense.

The Winter House

The winter house we built is a permanent, rugged structure that can withstand heavy snow cover, deep litter pack, and the use of hogs to dig the pack. We built it to be appealing—it's for our winter income makers and is the focal point of the farm. We also located it to allow access to power for frost-free water and ease of egg collection.

It's a wood-frame shed-roofed greenhouse off the west side of the barn, near power, water, and the egg-packing room. We used polycarbonate glazing sheets for their strength, clarity, and durability. The side panels are easily removed for ventilation, bird moving, or litter pack removal. The afternoon sunshine can easily warm the henhouse 15 to 30 degrees. The natural light also helps keep the hens productive and emotionally content. The structure has withstood heavy snow loads and strong winds.

The Setup

The interior is set up with the birds' natural behaviors in mind. The floor is dirt with a deep litter pack (12–36 in.) of softwood shavings, wood chips, and manure. There are 150 linear feet of roost space (0.5 ft./bird) in three tiers along the east wall of the barn. There are five 5-foot feeders down the middle of the house with two drinker founts on heating bases on the west wall. Also along the west wall are three self-feeders with supplements (grit, oyster shell, and kelp). The sixty nest boxes are located near the door from the barn in the southwest corner

The winter house.

for ease of collection. They are mounted on the walls 2 feet off the ground so as not to lose any floor space, and they are filled with softwood shavings (hardwood can leave stains on the eggs). There are also three hay nets filled with organic alfalfa hanging from the ceiling at the height of the birds' backs. Dusting areas are located at each end of the house. They are simply areas of dry shavings supplemented with wood ash, diatomaceous earth, and lime. I locate the big dusting area in front of the nest boxes so that area is always dry and clean before the hens step into the boxes. Having the boxes up off the floor protects the eggs from the fallout when the hens dust down below.

We supplement the natural light from the clear roof and three walls with two 100-watt bulbs to ensure a day length of 14–16 hrs. Winter production is economically important to us, so we consider additional light worthwhile to keep up egg laying in the winter. The lights are on a timer and go on at roughly 2:00 a.m. and shut off at 4:00 p.m.—before dusk so the hens can settle down as it grows dark and be roosting by complete dark. It is unfair to turn lights off suddenly and expect them to find their roosts in the dark.

The Deep Litter Pack

Management of the pack in a winter house can be challenging. It's a constant battle to maintain the proper carbon-to-nitrogen ratio (i.e., wood fiber to manure), particularly when you have the maximum number of hens on the pack. We prefer to start with a base foot of good composting litter from the previous year and add another foot of clean dry softwood shavings. Fresh sawdust, being fine and damp, doesn't absorb moisture very well and requires hand-turning with a fork at least weekly, because a pack that's not dry or loose enough to absorb the manure "caps" over in high-traffic areas (under the roosts) keeping the hens from scratching in it.

When the pack is light and well mixed, it is actively composting, reducing ammonia, and producing heat. The hens love to create bowls in this warm pack to lounge around in. We also occasionally add shavings where needed throughout the winter. In spring, after the hens move out to the summer hoop house, we remove half the pack and compost it further in windrows for future use on the perennial herb beds.

Winter Outdoor Access

Seasonal, carefully managed outdoor access is essential to an organic poultry system. Winter access, however, can be problematic in some climates—useless and even harmful if not properly managed. I believe hens are better off in a naturally lit barn on deep litter than wallowing outside in muck. This said, I do keep a small door open onto a large fenced paddock throughout the winter just in case they wish to go out. They don't. As spring rolls around and they grow more apt to venture outdoors, I monitor their access. Having the hens out in muddy conditions adds substantially to egg-cleaning time without benefiting the hens that much. Instead, I bring them fresh chickweed and other spring greens to eat and chunks of sod to scratch at. They will be out on grass soon enough when the pastures start growing. No matter where you stand on the outdoor-access question, you have to consider conditions and what is best for your hens and for the land and groundwater affected.

Winter Egg Collection

A primary purpose for keeping the hens close to the barn during the winter is to facilitate frequent egg collection. Depending on the weather, we typically collect eggs three times daily. If it is very cold in the hen house (below 10°F) we may collect more often, but this rarely happens since the hens, the litter pack, and the sun help keep the house warmer than outside temperatures. We collect directly into 30-hole plastic egg flats and store them in our egg cooler in an insulated room in the barn. The boxes are bedded with ample softwood shavings and cleaned out every morning. We also add a small amount of cedar shavings to help deter lice. We do not recommend hardwood shavings since they can stain eggs if they become wet and the stains cannot be removed.

Owing to cold nights in the winter some hens like to sleep in the nest boxes. We attempt to remedy this by closing up the top boxes at night (they like those the best) and removing any sleeping hens at the last night check. For clean eggs we find it is important to break hens of this habit as soon as it is noticed.

What Equipment and How Much?

Many types of feeders, drinkers, nest boxes, and other housing equipment are appropriate for organic poultry production (see appendices 1 and 2 for equipment sources). Although the current U.S. organic standards do not provide guidance for the feeder, drinker, roost, or nest-box space needed for poultry, it behooves the organic producer to carefully consider proper capacity of all equipment for the well-being of the birds. In order to use this equipment to the benefit of the birds certain ideals should be kept in mind.

- Feeders and drinkers should be labor efficient but not so large that you needn't check on the birds frequently.
- Birds should have enough feeder space to allow all birds to eat simultaneously unless you are providing continuous free-choice feeding.
- Drinkers should be kept full, thawed in cold weather, cool in hot weather, and free of contamination.
- Roost space should be sufficient for all birds (layers, turkeys, and guineas) to roost simultaneously.
- There should be a sufficient number of nest boxes or nesting areas so birds are not competing for space or breaking and soiling eggs.
- Dust baths or boxes must always be available to birds for the exercise of natural behavior and the control of external parasites.

The British Columbia certification program's recommended equipment capacities are shown in table 4 in the next section.

Outdoor Access

Under section 205.239 of the USDA's National Organic Program Regulation (NORP) Final Rule a producer of organic livestock must establish access to the outdoors, shade, shelter, exercise areas, fresh air, and direct sunlight suitable to the species, its stage of production, the climate, and the environment.

Providing birds with an outdoor environment is probably the biggest debate in the organic poultry industry. I say "industry," because among smaller producers this is mostly a non-issue. It is a general belief among small organic growers that access to the outdoors can and should be provided to all types of poultry in all stages of life. Pasture-based systems are widely used in Europe on a large scale and have been shown to be both economically and environmentally viable. However, many large producers in this country have contended that outdoor access at a mini-

Table 4. Recommended Equipment Capacities	
Brooding, all poultry	
Feeder space	5 ft. of feed trough/100 birds for first 4 wks. 2 bucket feeders/100 birds for first 4 wks.
Water space	1.5 round 5-gal. waterers/100 birds for first 4 wks.
Laying or breeding fowl	
Roosting space	8 in./bird
Feeder space	1 ft. of feed line/4 birds 1 tube or pan feeder/15 birds in controlled feeding environment 1 tube or pan feeder/40 birds in a free feeding environment
Water space	1 ft. of water trough/15 birds 1 round waterer/100 birds
Nest space	1 well-maintained nest box/4 laying hens
Meat birds*	
Feeder space	1 ft. of feed line/10 birds 1 tube or pan feeder/30 birds
Water space	4 ft. of trough waterers/100 birds 1 tube waterer/100 birds

Source: British Columbia Certification Program.
*These standards give no specifics for ducks, geese, turkeys, or other fowl.

mum is unhealthy for birds and that pasture access is completely unviable. As evidenced by the complaints brought by the Country Hen against its certifier, which required outdoor access beyond covered porches, some large growers want to adapt the new organic rules to their existing systems, which historically had been accepted by some certifiers.[8] In May of 2002 the National Organic Standards Board (NOSB) voted to keep the outdoor-access rule intact, but its position hasn't been adopted as of this writing.[9]

The author's Kingbird Farm pullet house.

What is outdoor access and why is it important? Not only the NOSB, but the Humane Society of the United States (HSUS) issued its own recommendations based on the work of M. C. Appleby, who carried out research on behavior, housing, and welfare of poultry for twenty years at the University of Edinburgh, a bastion of livestock-welfare research. The HSUS voiced support for all four principles that the NOSB had listed as the intent of the outdoor-access recommendation:

1. To satisfy poultry's natural behavior patterns (i.e., foraging, dusting, and exploration).
2. To provide adequate exercise area (foot condition and wing and leg-bone strength).
3. To provide preventative health-care benefits (reduction of stress, strengthened immunity, and varied nutrition).
4. To answer consumer expectations of organic livestock management.[10]

There are several arguments against outdoor access made by large producers trying to adapt to the new federal rule. Many large, barn-based operations that were certified organic prior to the implementation of the federal rule were large open aviary-style setups that provided deep litter, dusting areas, roosts, and natural light and ventilation, but no outdoor access. Providing such large concentrations of birds with any sort of viable range may be nearly impossible. Yarding and dirt runs often create "fowl-sick" land and a situation that may actually be inferior to confining the birds to deep-litter floors. During discussion on changing the definition of outdoor access in the final rule, North Carolina state veterinarian David Marshall declared that "while noble in its intent, this concept (outdoor access) is ill conceived and not conducive to mass rearing of poultry."[11] Indeed, he may be correct that organic production is no place for large confinement operations. Perhaps "large" organic growers can only achieve mass productivity by contracting out to several small farms that are willing and able to follow not only the letter of the law on organic management, but also its spirit.

Beyond Outdoor Access: Grazing and Pasturing

Besides mere outdoor access, there is a strong tradition among organic producers to provide poultry of all types with high-quality forage on range. Proper forage is part of the bird's natural eating behavior and diet, and many would say it adds flavor to their eggs and meat. It certainly increases a key nutrient content—it has been known for years that eggs from birds raised on actual forage are higher in omega-3 fatty acids and other nutrients[12] and current studies continue to confirm this in birds raised on managed, rotated grass/legume pastures, not those with dirt yards or sun porches.[13] (Not all nutrients have been similarly compared.)

As monogastric animals with a small cecum (the organ that processes roughage), chickens are not designed to utilize forage in large amounts, but they can make use of high-quality forages, particularly legumes. Useful forages include Ladino clover, Sudan grass, oats, wheat, and alfalfa plus many other grasses and broadleaf plants. There are typically more unsaturated fatty acids in legumes, particularly the leafy portion of the plant.[14] Many grass-based poultry farmers have developed favorite forage mixes, which they seed into their pastures or simply encourage to thrive with proper liming and mowing. The forage types chosen will vary according to the type of poultry and other livestock being grazed. While geese and

An "eggmobile," or movable-style range shelter.

older turkeys can actually be grazed on grasses, chickens and ducks are better suited to foraging legumes and other broadleaf plants.[15]

Actual utilization of pasture by various types of poultry varies from 100 percent of needed nutrition by geese to possibly 30 percent by chickens. Producers and researchers have estimated that forage constitutes anywhere from 5 to 30 percent of a pastured chicken's diet.[16] The high end of this estimate is rarely seen, and many pastured-chicken producers actually find an increase in grain consumption regardless of forage intake. The amount of forage actually utilized by birds may vary considerably according to forage quality, breed of bird, and amount of grain ration provided. Also, birds cannot be expected to fully utilize fibrous pasture plants without being provided with grit. In light of all the above factors, most producers choose to keep a diverse, perennial mix of forages that are managed by grazing (by ruminants) or mowing to maintain leafy growth.

One recommended pasture mix used by producer Herman Beck-Chenoweth[17] is as follows:

½–1 lb. ladino clover
6 lbs. medium red clover
2 lbs. alsike clover (for wet soils)
6–8 lbs. brome grass

Seed mixes should be formulated according to a variety of soil and agronomic factors including other grazed livestock using the pastures.

Managing any type of poultry on pasture requires proper fencing and access to feed, water, and shelter. The ATTRA publication "Sustainable Poultry Production" (available online) provides a good discussion of fencing types and feeder and waterer design. All these systems are adaptable to organic free-range pasturing.

three

Purchasing and Brooding Chicks

Currently the organic production of poultry requires the use of day-old chicks that have been managed organically from their arrival on-farm. Unlike in the UK, organic growers in the United States are not required to purchase certified organic chicks. As a consequence there are currently no commercially available organic chicks, poults, goslings, or ducklings. Organic producers in the United States are also surprisingly limited in their choice of genetics to match with their type of system. Vertically integrated poultry companies primarily control the development of breeding stock in this country. Most of the independent hatcheries that organic producers deal with do not keep their own flocks to produce hatching eggs. Therefore the majority of chicks available from hatcheries throughout North America are the same strains used by commercial producers. There has recently been some interest in the development of new strains.

Meat Chickens

Organic producers use the Cornish Cross chicken not because it is ideally suited to rearing on pasture, but because it is readily available. No other type of chicken widely available in North America produces as much meat as economically as the Cornish Crosses. Some hatcheries offer slower-growing strains of Cornish Crosses, which may be useful to free-range organic growers because they may have fewer leg and heart problems. A publication by Anne Fanatico and Skip Polson, "Which Bird Shall I Raise?" covers the development and availability of alternative breeds for broiler production.[1] Some currently available options for meat birds are listed in table 5.

Table 5. Meat Chicken Birds

Breed	Characteristics	Availability (see appendix 3)	Average price per 100 chicks
Jumbo White Cornish Cross × Rock White Cornish Super Giants, Fast Cornish, Vantress × Arbor Cross, Hubbard White Mountain	Fast growing, heavy-breasted bird with excellent feed efficiency. Prone to leg problems, ascites (water belly possibly caused by inadequate development of the respiratory system),* and flip over (sudden death syndrome/heart attack, possibly caused by rapid weight gain).**	Reich Hoffman Marti Welp's Privett Mt. Healthy Clearview Moyer's Townline	$0.63-$0.95 each
Barred Silver Cockerels	Slightly slower growing dark birds that can be held longer for roasters; excellent foragers with few health problems.	Reich Hoffman Ridgeway Clearview Noll's	$0.37-$0.69 each
Buff Silver Cockerels	A lighter colored slower growing bird similar to the Barred Silver; same growth rate with outstanding leg strength and vigor.	Reich	$0.50 each
Black Broiler Cebe Black	Black birds that are active, slow growing, and with few leg problems or ascites.	Welp's Privett Cebe's Townline	$0.75-$1.13 each
Red Broiler Red Cornish, Cebe Red	A red bird that grows slower and smaller than the standard Fast Cornish.	Welp's Privett Clearview Cebe's	$0.98-$1.13 each
Slow Cornish	A white bird that grows slower and smaller than the fast Cornish; they have fewer growing problems.	Privett	$0.89 each

*Gail Damerow, *The Chicken Health Handbook* (North Adams, MA: Storey Publishing, 1994).
**Ibid.

Laying Hens

The range of egg breeds available is extensive, though not specifically geared toward commercial free-range organic systems. Breeds available range from old heritage dual-purpose birds with relatively low production to modern hybrids bred to produce a maximum number of eggs in

Table 6. Laying Hen Breeds

Breed	Characteristics	Availability (see appendix 3)	Average price per 100 chicks
White Leghorn (white eggs)	A small, nervous, refined bird requiring a high-quality diet and controlled environment.	Reich, Clearview, Hoffman, Mt. Healthy, Privett, Welp's, Marti, Moyer's	$0.97-$1.50 each pullet
Gold Sex-Link Golden Comet, Production Red Sex Link, Hubbard ISA Brown	Rhode Island Red/White Rock cross. Medium-sized brown bird that forages well and produces in cold weather. Tends to cannibalism if stressed. Most efficient brown egg layer with rates of 90%.	Reich, Clearview, Hoffman, Mt. Healthy, Privett, Welp's, Marti, Moyer's	$0..97-$1.50 each pullet
Black Sex-Link Black Star	A Rhode Island Red/ Barred Rock cross. A medium-large black bird capable of commercial production.	Reich, Clearview, Hoffman, Mt. Healthy, Privett, Welp's, Marti, Moyer's, Townline	$0.97-$1.50 each pullet
Rhode Island Red	A medium-sized docile native purebred with good foraging behavior. Capable of 70% rate of lay.	Reich, Clearview, Hoffman, Mt. Healthy, Privett, Welp's, Marti, Townline	$0.97-$1.52 each pullet
Barred Plymouth Rock	A larger-sized native dual-purpose breed with dark feathers. Good foragers with a 60% rate of lay. Cold weather layer.	Reich, Clearview, Hoffman, Mt. Healthy, Privett, Welp's, Marti, Townline	$0.97-$1.50 each pullet
White Rock	A large, fast-maturing dual-purpose bird with decent production.	Ridgeway, Mt. Healthy, Privett, Welp's, Marti, Townline	$1.15-$1.46 each pullet
Production Red	Hybrid layer cross developed from the RI Red as more efficient layer. Since they have brittle feathers and body size is small relative to the size of eggs they are susceptible to cannibalism.	Privett, Mt. Healthy, Marti	$1.14-$1.46 each pullet
Cherry Eggers	A heavy red bird with vigor and good livability. A decent producer that lays well in cold weather.	Marti	$1.17 each pullet
Sil-Go-Links	A smaller brown hybrid layer, good feed efficiency. Non-cannibalistic.	Marti	$1.17 each pullet
Silver Links	A large dual-purpose hybrid with fairly good feed efficiency and excellent production. Matures early and feathers well.	Marti	$1.17 each pullet

intensive systems. Organic production generally requires a bird somewhere between these two extremes. Most organic systems want a bird that can produce efficiently under a range of conditions and utilize forage to its best advantage. While organic producers in Europe have a vast array of free-range genetics to choose from, producers in the United States are generally limited to a few brown-egg-laying hybrids that produce efficiently under free-range conditions. See table 6 for more on laying hens.

In addition to the commercial breeds listed in table 6 there are also myriad standard old-style breeds that some producers find to be excellent foragers but are perhaps less productive. In my survey of small organic producers of the Northeast, I found producers to be using a vast selection of both hybrid and standard breed layers. The top three breeds chosen were Rhode Island Reds, Barred Plymouth Rocks, and Ameraucanas. However, among the producers with 200 or more hens, the top choices also included hybrid layers such as Black Sex Links, Gold Sex Links, Red Stars, and ISA Browns. Most flocks were mixed breeds and only one producer reported using just one breed. On the other end of the spectrum, large commercial organic growers tend to rely on one breed of hybrid layer that provides consistent and economical production. Producers cannot feed certified organic feed to large flocks of poorly producing birds and remain economically viable, no matter what kind of system they are using.

Turkeys

Turkey genetics in this country are basically twofold: large white heavy-breasted quick growing commercial birds and approximately ten heritage breeds with varying size, color, and grow-out time. Though most organic producers currently use the standard broad-breasted white turkey owing to its availability and efficient grow out, there has recently been much interest and research into heritage breeds. SARE is currently funding a study to determine biological fitness and productivity in range-based systems comparing standard turkey varieties (heritage breeds) and industrial stocks.[2] This study is designed to provide producers with information about the superior health and survivability of standard turkeys in

Table 7. Turkey Breeds			
Breed	**Characteristics**	**Availability (see appendix 3)**	**Average price per 100 chicks**
Broad Breasted White, Large Broadwhite	A very large, white, fast-growing bird with excellent feed conversion. Tom, 24 lbs. at 24 weeks; hen, 17/24. Cannot reproduce naturally. Good foraging once fully feathered.	Reich, Welp's, Hoffman, Mt. Healthy, Privett, Ridgeway, Townline	$2.00-$4.65 each poult
Giant Bronze, Broad-Breasted Bronze	A large, dark, fast-growing bird with good feed conversion. Cannot reproduce naturally.	Reich, Welp's, Hoffman, Mt. Healthy, Privett, Ridgeway	$2.75-$4.80 each poult
Beltsville White	White. Young tom weighs 17 lbs. at 6 mos.	Welp's, Hoffman, Privett, Ridgeway, Walters	
Black	Black. Young tom weighs 23 lbs. at 6 mos. Good eating bird, but hard to remove black pinfeathers.	Welp's, Hoffman, Privett, Ridgeway, Walters	$7.45-$9.20 each poult
Blue Slate	Ashy blue. Young tom weighs 23 lbs. at 6 mos. Good eating bird that does not roam far.	Welp's, Hoffman, Privett, Ridgeway, Walters	$7.45-$9.00 each poult
Bourbon Red	White and red. Young tom weighs 23 lbs. at 6 mos. Production type conformation with a heavy breast. Good forager.	Welp's, Hoffman, Privett, Ridgeway, Walters	$7.45-$9.20 each poult
Bronze, Standard	Copper bronze and black. Young tom weights 25 lbs. at 6 mos.	Privett	$7.45 each poult
Jersey Buff	Reddish buff and white. Young tom weighs 21 lbs. at 6 mos.		
Narragansett	Black, gray, tan, and white. Young tom weighs 23 lbs. at 6 mos. Needs proper nutrition to attain weight.	Welp's, Privett, Walters	$7.45-$8.00
Royal Palm	White with black edging. Young tom weighs 16 lbs. at 6 mos. Not a meat bird.	Walters	$6.00-$9.20 each poult
White Holland	White. Young tom weighs 25 lbs. at 6 mos. Very gentle, easily managed, good meat bird	Walters	$10 each poult
White Midget	White. Young tom weighs 13.8 lbs. at 6 mos.		
Wild Turkey (captive bred)	Very hardy bird with minimal meat. Flighty and requires netting.	Welp's, Hoffman, Privett, Ridgeway, Walters	$7.95-$9.20 each poult

range-based systems. Since organic turkey poults must be reared without the use of antibiotics, the natural immunity and healthy vigor of heritage turkeys may be of interest to organic growers. These birds are also known to forage well and produce excellent flavor over their long grow-out time. However, these breeds do have disadvantages, particularly to the organic grower who may have to put a substantial amount of organic feed and time into this specialty bird. See table 7 for more on turkey breeds and their availability.

Waterfowl and Other Poultry Breeds

There is little information about the types of breeds that organic producers are using for rearing waterfowl and other poultry. Commercial breeds of waterfowl such as the Pekin duck and Emden goose and most other waterfowl breeds are suited for free-range organic production. Geese do well on pasture since they are true grazers and can make excellent use of pasturage. Most breeds of duck would also be suitable for organic production whether for eggs or meat. As with chickens and turkeys, two important factors to consider when choosing breeds for organic production are good feed conversion and adaptability to free-range conditions.

Special Problems Brooding Organic Chicks

The brooding of organic chicks is identical in many ways to conventional brooding with a few key differences. Although it is a commonly held belief that chicks and poults cannot be started without coccidiostats and other antibiotics, organic producers have been doing just that with excellent results. Many smaller producers (fewer than 1,000 birds) use a combination of immunity-enhancing techniques such as apple cider vinegar, deep litter pack, soil contact, fresh forage, and excellent nutrition. In my survey of smaller producers very few reported any problems with coccidiosis or other brooder diseases, though large commercial flocks may encounter this disease more regularly due to stocking densities and associated stress.

Many organic producers also brood directly on pasture or provide young chicks with access to sun porches or small runs. This technique helps acclimate young birds to the many environmental factors they will face as adults on range. Early access to the outdoors accustoms chicks to natural light cycles, weather patterns, predator risks, and foraging opportunities so they can better thrive and produce under these conditions. However, it is not advisable to move young birds out to pasture or range too early or in inclement weather just for the sake of getting them out. If wet, cold, or dangerous conditions prevail outdoors, it is always preferable to continue their confinement to the brooder until the transition is appropriate. Perhaps keeping young birds housed and providing them with harvested forage or sod during harsh weather is a workable solution.

THE AUTHOR'S EXPERIENCE:
Organic Brooder Management[3]

As spring rolls into central New York, it is time to ready the brooder for chicks.

Most baby poultry we order, with the exception of Shell's broilers and the turkeys, are readily available and can be ordered nearly any time. If broiler chicks are ordered from large hatcheries like Reich's, there really isn't much of a rush either. There is, however, only a small window for hatching turkeys, and if you need them on a certain date (especially a later date in July) order them early. I like to have all my ordering done in February so I know when chicks are coming and can schedule brooder availability and processing dates.

All hatcheries require a minimum order of 25 chicks. There are four sections in a standard shipping box; each section holds 25 chicks. This number is necessary for the chicks to maintain body heat during shipment. Chicks from Pennsylvania hatcheries usually come by ground and those from farther away are shipped by next-day air. In my area, chicks arrive by truck at my local post office, which then calls me when they arrive. If you are planning to receive hundreds or thousands of chicks at your post office, it behooves you to make it easy on the postal staff. Arrive promptly when they call (chicks can be noisy in a small post office), and take them some eggs. While still at the post office, inspect the chicks and check for any dead. If you need to make a claim for dead chicks you must do it at the post office. When counting dead, remember the hatchery usually adds an additional two or three chicks per box. No matter how far away our chicks have come, we rarely have any death in shipment. The system works well.

The Brooding Structure

Prior to the chicks' arrival, the brooder house must be prepared for the season. My brooder is in use nearly year round and must be refitted for the chicks. The last batch of birds using the brooder in fall is replacement layers (100). These pullets spend all winter in the brooder on a deep pack accumulated throughout the season. As they begin laying in early March, I move them into the chicken greenhouse with the mature layers. This gives me a chance to renovate and repair the brooder before the broiler and layer chicks arrive the first week of April.

The brooder design is simple with a few key elements that make it work well for us. The structure is 10 × 12 ft. with a 12 ft. front and a metal shed roof. The floor and walls up to 3½ ft. are rough-cut poplar. From the low wall to the shed roof it is clear plastic to allow as much light as possible. The entire brooder is lined with 1-inch chicken wire, and the floor is covered with hardware cloth. Attached to the rear of the building is a 10 × 12 ft. plywood deck covered by an old Salatin-style pasture pen. The chicks access this "poultry patio" through a small door in the brooder wall.

We designed with the chicks in mind to prepare them for their lives out on pasture. The plastic walls allow abundant natural light, and the "poultry patio" gives them a chance to adapt to weather and foraging within the safety of the brooder environment. We can begin the transition to a pasture-based life by regulating their day with natural light and feeding them grasses and other plants on the patio. This method has reduced our pasture mortality to almost zero.

We prepare the interior of the brooder to accommodate 200 to 300 chicks for two to three weeks. The previous year's pack is removed and shoveled into the pig pens for further mixing and composting. We air out the brooder and check for holes. During the winter, under deep pack, the rats may have spent a lot of time figuring out how to gain entry. Any holes are covered with hardware cloth and torn plastic is replaced. The patio is also shoveled out and allowed to dry. We do not wash, sanitize or sterilize the brooder.

Once the brooder is sealed up again, we start a new pack with 6 to 8 in. of hardwood shavings that we buy in bulk from the local sawmill. We do not use sawdust (too dusty), cedar, walnut, or cherry shavings (risk of poisoning). Two 250-watt red heat lamps are suspended from the ceiling on the side of the brooder away from the door. Directly inside the door are two or three low 3 ft. trough feeders and two 1 gal. plastic drinkers placed on boards for stability on the shavings. As the chicks grow larger, feeders will be added, and a Plasson bell drinker with a 5 gal. bucket will be suspended from the ceiling. When the chicks first arrive we offer them chick grit and feed on old egg flats for easy access and later keep grit supplied in a metal feeder on the wall.

The Amenities

No matter how large your brooder is, it is vital to have it ready and waiting when the chicks arrive. Right after I hang up the phone with the post office I turn on the heat lamps and prepare warm water with molasses and apple cider vinegar. Warm water is particularly helpful to chicks that arrive in March. To each gallon of warm water I add 1 tsp. of organic blackstrap molasses for energy and 1 tbsp. of organic apple cider vinegar to increase the acidity of the chick's gut and help prevent coccidiosis. Needless to say, I do not add any coccidiostats. Using this water mix and a deep litter I have only had one brief bout of coccidiosis in layer chicks that resulted in the loss of 10 chicks out of 200.

When I place chicks in the brooder I make sure to dip their

beaks in the warm water. It's like priming them. Typically they will then huddle up under the lamps for a while before venturing out to peck feed and grit out of the egg flats. Within a few hours they are well distributed among the heat, feed, and water locations. I am careful to observe them several times during the first day to make sure there is no piling up, overheating, or other problems.

The chicks are confined to the brooder for the first week. After this, if the weather is mild, we begin allowing them onto the patio. They have already been receiving chickweed, plantain, and clipped grass in the brooder, and now it is time to begin foraging through chunks of sod and bunches of plants (clover, chickweed, dandelion, and vetch) on the sunny patio. At this time we may also begin turning off the heat lamps if it warms up during the day and keeping them off at night as well. After another week or two, depending on the weather, we gather them up in crates and move them out to pasture.

Feedstuffs

The feed that chicks (especially broilers) receive in the brooder is critical to their start in life. It must fulfill all their nutritional requirements and support their immune systems. Our feed is a custom-ground certified organic grower mixed with select supplements and locally grown grains whenever possible. It is ground, mixed, and delivered bulk by Lakeview Organic Feeds in Penn Yan, New York. Originally it was based on Joel Salatin's recipe, but grain availability and certification regulations have forced us to adapt it. Previous NOFA-NY standards did not allow for the use of crab or fish meal, which supplied essential amino acids for quickly growing broilers. Chickens are not vegetarians, and growing chicks cannot find enough animal protein on pasture to fulfill their needs. Without access to any animal proteins we were forced to add a synthetic amino acid (dl-methionine) to our grower feed in order to keep broilers and turkeys from falling apart. With the new national standard we can use crab meal to complete our ration.

Pests and Predators in the Brooder

Protection of very young chicks in the brooder is difficult but critical. For many years we did not have predator problems in our brooder—perhaps an occasional mouse or circling hawk, but no predation. That all changed one recent year when the rats moved in. They found refuge under our barn's cement slab and discovered a cornucopia of organic chicken feed to pilfer. All year we battled them with traps, cats, dogs, chicken wire, pitchforks, and even approved bait (Quintox). But nowhere was the war harder fought than in the brooder. Rats love to snack on baby chicks. They flay the little chicks expertly or simply eat their heads and leave the rest.

We lined the brooder with chicken wire, so they ate through the floor. We put bait behind the wire and they ignored it (why eat it with so much yummy chicken feed around!). We installed high decibel alarms, so they got earplugs (we think). They found every crack and crevice we failed to seal and slipped in every night to gorge. They particularly like the tiny Silkie chicks. Once broiler chicks were a week old the rats pretty much stopped bothering them, but before that age, the chicks were in danger. We continue to fight these battles. The war is far from over, but we keep the wily rodents to a dull roar. We've even considered raising a brooder cat.

Deep Thought on Deep Litter

One of the biggest differences between our brooder management and that of conventional operations is the deep litter pack. We do not remove litter at any time during the brooding season. After each batch of chicks we add 6 to 8 inches of fresh shavings on top of the pack. This is continued throughout the season until the last batch of layer chicks takes over occupancy. These birds remain on that pack all winter while it composts. The litter, now compacted down to approximately 1½ ft., heats up and warms the brooder all winter. The young hens love to dig holes in the pack and dust in the warmth.

Leaving this pack intact jump-starts the manure composting process and warms the birds at the same time. In addition, we believe it provides an incubator for good bacteria and a mild immune booster for the chicks. Each batch of chicks is lightly exposed to the bugs of the last batch, giving them natural immunity and vigor. In our six years of using this method we have not had any major illness. We get a little pasted vent early on which we think is due to the chicks not drinking enough water when it's cold. Last year we had our first small case of coccidiosis when a new batch of Rhode Island Reds arrived in the brooder already housing two-week-old Dominique chicks. The Reds immediately came down with coccidiosis and started dropping. I continued to add vinegar to their water and let the illness take its course. Five Reds died within the first couple of days and then five more subsequently. The older Dominiques never showed any signs of sickness. Had I treated with coccidiostats I would have still lost that many and wouldn't, I feel, have strengthened the immunity of the other chicks. Our brooder losses usually occur within the first day (just weak chicks) or from rats.

My organic brooder-management approach is based on the principles of natural light, wholesome feed, outdoor access, and greens to eat. If the brooder is a warm, bright, good smelling environment where you might like to curl up for a nap, then it's right for your chicks too.

---

four

Organic Feed and Supplements

In organic egg or meat production, feed will typically be the major expense. Jacquie Jacob, a poultry nutritionist at the University of Minnesota, says the cost of organic feed amounts to roughly two-thirds of the final production costs.[1] Producers often end up using whatever feed they can get their hands on regardless of quality or price. Much of this problem is due to regional access and the scarcity of certified organic mills. A USDA study of organic feed availability and prices concluded that there is ample organic acreage available in the United States to provide feed grains to meet the needs of organic *broiler* producers. It also found that organic poultry rations are *not* typically more than twice the price of conventional poultry rations.[2] The challenge for most producers will be obtaining this feed locally and having it ground and formulated according to their needs.

Basic Nutrition for Poultry

The laying bird does not know she is producing organic eggs and the meat bird does not know he is producing organic meat. It's the environment that mainly influences their nutritional needs, not the fact that they are part of an organic production system. So from a nutritional point of view the diet should be similar to one designed for any free-range system. A good jumping-off place for basic recommendations is Jeff Mattocks' publication "Pasture-Raised Poultry Nutrition."[3] In this publication he provides detailed specific nutritional requirements for layers, broilers, ducks, and turkeys and what feedstuffs might be used to fulfill these requirements.

Buying Bagged Feed

Many small producers might have only a few feed options available to them due to locality or amount they are willing to buy. Bagged feed is more expensive than bulk owing to production and packaging, but is formulated for the proper nutrition of the type of poultry it is designed for. Some mills will bag a custom mix for orders of more than a ton (50 bags). If buying bagged feed is the only option, it is important to note the freshness of the mix and whether it is certified by an accredited certifier. For farms that are not certified organic it may be more economical to purchase transitional feed containing grains that are produced by farms working toward the organic certification of their land. It is important that these farms have a market for their grains to encourage the transition of conventional land to organic management.

Developing an Organic Feed Recipe

When formulating rations suitable for organic production, farmers must consider not only the nutritional requirements of the birds, but also requirements set down by the NOP. All ingredients, supplements, and additives in feed must be reviewed before they can be used in certified organic production. Some formulas developed for free-range systems cannot be directly translated into certified rations because of the restriction or prohibition of certain ingredients. Due to the current prohibition of fish meal (or more specifically the preservative it contains) and the phaseout of dl methionine in 2005 it may become increasingly difficult to formulate an organic broiler/turkey feed with the proper amino acids. The amino acids in question are found in substantial quantities only in animal products, most of which are banned from use in organic feeds.[4] While many organic chickens and eggs are touted as fed only a vegetarian diet, most of these diets do contain the synthetic amino acid dl methionine to make up for the lack of animal proteins. While birds, like geese, can graze for a living, chickens and turkeys are not vegetarians and require the proper balance of amino acids to complete their protein intake. It is also important to keep in mind that the digestive system of the chicken is geared toward the digestion of insects, seeds, and grain rather than forage, and they need concentrates. A commercially viable chicken, turkey, or duck flock cannot be sustained on range alone.

Despite the above challenges, some producers have successfully formulated their own feed rations, typically with the help of a poultry nutritionist like Jeff Mattocks or the livestock nutritionist at their mill. A nutritionist can formulate rations based on which certified feed grains are available to your feed mill, which supplements are allowed, and exactly how the birds are going to be raised. Many types of poultry require differing amounts of protein depending on their stage of growth.

Adjusting this protein percentage can produce not only steadier, more natural weight gain, but also decreased feed costs. There is no point in providing a laying hen with 22 percent protein; it might adversely affect her production. Some small producers attempt to use a single feed for all livestock, adjusting the protein by adding oats or changing the formulation by mixing in a vitamin pack depending on the type of animal being fed. This method works only if it's carefully monitored and could eventually cause problems with fast-growing birds like Cornish Cross broilers or nutritionally sensitive young turkeys. Birds like pullets need to be fed for a slow rate of gain that is intentionally designed to allow the bird to develop a stronger metabolism and immune system and reduce fat production.[5] Therefore it is inadvisable to raise pullet chicks on the same feed as broiler chicks then abruptly switch them to a laying ration. Feeding the wrong ration can also decrease the rate of lay and make hens fat. When the price of organic feed is high and the value of the product is high, it is more than worthwhile for a producer to choose feeds that maximize production.

Working with Feed Mills

When buying organic feed directly from the mill, either bagged or bulk, a producer should consider the following factors:

Price

- Bagged feed is more expensive due to production, packaging, and labeling costs.
- Bulk feed is typically less expensive but requires the installation of bulk storage.
- Formulations may need to be adjusted if certain ingredients like soybeans become prohibitively expensive.

Convenience

- Large-scale mills will probably require at least a 3-ton order for custom grinding or delivery.
- Small-scale mills will probably custom mix 500 lbs.-1 ton, but may require a much larger order for delivery.
- Bulk feed can be gravity fed or augured.
- To save investment in trucking and storage systems, smaller operations may prefer bagged feed.
- Most small mills do not have pellet or crumbles capability and usually only grind mash.

Freshness

- Buy in bulk only if the feed will be used within 45 days.
- Oxidation of the feed starts immediately after grinding or cracking and freshness is optimum for up to 14 days and satisfactory up to 45 days.
- After 45 days feed is generally so stale or oxidized that it will decrease a bird's appetite.
- Airtight containers will extend storage.

Feeding Strategies

Before deciding on a feeding strategy an organic poultry producer should review some general guidelines. Table 8 will help plan feed purchases and storage. Amounts of feed are determined by many factors including, but not limited to, weather, outdoor access, forage quality, stage of production, bird behavior, feeder space, and whether other supplemental feed is available. Limited as it is, the information in the table will give the producer an idea of what feed consumption to expect for conventionally raised birds in order to make initial feed orders. Keep in mind that free-range birds will commonly consume considerably more feed than confined birds due to exercise, weather, and other factors. The chart is not intended as a guide for performance or other management practices.

Once it is determined roughly how much feed a bird can be expected to consume, it is important to develop a feeding method that optimizes

Table 8. Feed Consumption (Pounds/Bird Cumulative Feed Consumed)					
Age	Meat chickens	Egg chickens	Turkeys	Ducks	Geese
0–4 wks.	2.3	1.8	2.1	6.3	6.5
To 8	8.6	4.6	8.8	20.0	22.0
To 12	18.1	8.4	25.0	33.0	35.0
To 16	30.8	12.0	44.0		
To 20	44.0	17.6	63.0		
To 28	–	–	116.0		
Layers 0.2– 0.25 lb./bird/day					
Source: Mahmoud El-Begearmi, Amounts to Feed Your Flock, Poultry Facts, Bulletin #2051, University of Maine Cooperative Extension, 2001.					

both production and health of the bird. There are three basic methods of offering feed to birds, each having its advantages and disadvantages according to the type of bird and the production system being used:

Continuous free-choice feeding

- Allows access to a full ration at all times.
- Allows birds to regulate their own feed intake.
- Prevents crowding and fighting during feeding.
- May result in sorting if feed is not pelleted.
- May result in feed waste and contamination if feeders are not properly designed.
- May attract more wild birds and rodents to feeders.
- May cause layers to become fat and broilers to grow too quickly.
- May require moistening feed with water, flax oil, or soy oil to increase palatability and uptake of fine particles if sorting is occurring.

Controlled feeding

- By timing and ration control, limits birds to amount that they can consume before the next feeding.
- Allows farmer to moderate and control the bird's feed intake.
- Reduces sorting.
- Encourages birds to forage between meals.

- Requires ample feeder space to prevent crowding and fighting during feeding.
- Allows farmer to use feeding as a tool to move birds to another location.
- May prevent smaller, less aggressive birds from getting enough to eat.
- May cause picking in younger birds looking for something to do.

Self-selection or cafeteria-style feeding

- Allows a choice of whole grains or other feeds along with a full ration.
- Allows birds to self-regulate their nutritional needs.
- May result in considerable feed cost savings.
- Requires a constant supply of grit available to process grains.
- Is not appropriate for young, fast-growing birds.
- May result in birds ignoring full ration and missing out on needed supplements.

Feeds offered for self-selection or cafeteria-style feeding alongside complete rations could include the following:

- Chickens—wheat, sprouted oats, corn (winter), alfalfa.
- Turkeys—whole wheat, oats (to slow growth).
- Ducks/geese—cracked or whole corn.
- All types—vegetable scraps and weeds, but only as much as they will clean up in ten minutes.
- Avoid strong-smelling vegetables or herbs that may taint eggs and meat.

No matter what type of feeding system a producer chooses, it is important to remain aware of how birds are performing and the quality of their health. Sometimes the easiest method is not the best for the birds or the economic viability of the farm. Robert Plamondon, a free-range producer in Oregon, once made a very good point about how we tend to feed layers—one that's particularly poignant for small producers who use heritage breeds and find it difficult to make them economically productive. He points out: "A laying hen is a breeding animal, not a production animal. She is in the same class as the cow. Laying hens are essentially giving birth daily. For optimum results they should be fed like breeding stock and receive a daily portion, and not like production stock with the feed trough running over . . . you need to handle the hens and feel how much flesh and fat they are carrying. Barred Rocks will eat themselves into the stockpot. Rhode Island Reds are not quite as prone, and Leghorns are not capable of this at all."[6]

five

Poultry Health Care in the Organic System

Section 205.238 of the National Organic Program Final Rule (see appendix 7 for the complete rule) states that the producer must establish and maintain preventative livestock health-care practices, including:

1. Selection of species and types of livestock with regard to suitability for site-specific conditions and resistance to prevalent diseases and parasites;
2. Provision of a feed ration sufficient to meet nutritional requirements, including vitamins, minerals, protein and/or amino acids, fatty acids, energy sources and fiber (ruminants);
3. Establishment of appropriate housing, pasture conditions and sanitation practices to minimize the occurrence and spread of diseases and parasites;
4. Provision of conditions which allow for exercise, freedom of movement and reduction of stress appropriate to the species;
5. Performance of alterations as needed to promote the animal's welfare and in a manner that minimizes pain and stress; and
6. Administration of vaccines and other veterinary biologics.

Organic poultry health care is based primarily in prevention, and this manual has already covered most of the above aspects. However, since organic poultry growers generally regard the perceived need for "alterations" like debeaking or beak trimming as symptoms of a problem rather than preventative measures, it will not be covered here as a part of health care. Another issue that the U.S. standards fail to recognize is the development of a bird's natural immunity and its ability to partially manage its own health care if provided with the means to do so. Organic poultry health care is a holistic affair that recognizes health as a state of equilibrium among many factors, not least of which is the human factor. The holistic approach must start with the farmer. The following list of guiding principles for maintaining poultry health is adapted from Taylor Hyde:

1. Adjust your attitude; think and feel that *you* are the cause of all disease in your flock. This prevents "attitudeosis"—nothing changes until your attitude does.
2. Put poultry only in places where you would want to be. This will prevent stress and immune system suppression.
3. Offer only water you would drink. This prevents dehydration and the spreading of germs.
4. Do not starve or overfeed. This prevents growth problems.
5. Offer a diversity of feed and forage. This prevents nutritional imbalances.
6. Allow poultry only on ground with a minimum amount of manure. This minimizes parasite exposure, coccidiosis, and diarrhea.
7. Avoid sudden feed changes. This helps prevent diarrhea, crop impaction, pasted vent, and picking.
8. Cull. This prevents "wimp" birds from threatening the health of other birds.
9. Handle and move birds gently. This prevents stress, injuries, and fearfulness.[1]

The Role of Vaccinations in Organic Production

Vaccines have risks as well as benefits for the organic poultry producer. Generally, commercial flocks over 5,000 are vaccinated for a number of common diseases, but smaller flocks, particularly those raised on pasture, typically do not need vaccines unless a certain disease is common in the

Based primarily in prevention, the holistic approach to poultry health care starts with the farmer. Provide a clean environment, only water you would drink, and disperse feed consistently and in the right amounts.

area.[2] Vaccinating for Marek's disease, a contagious litter-borne disease caused by herpes viruses, is commonly recommended for layers since they are kept so much longer than broilers.[3] It seems likely, however, that this preemptive measure does not allow the bird's own immune system to naturally generate its own defense.[4] This shortcoming may hold particularly true in brooders that are sterilized between batches. Rearing birds in a clean, hygienic environment is good in many ways, but the young birds may not meet a wide enough range of infectious agents and get the opportunity to develop a broad immunity or perhaps a fully functioning immune system. This might leave young birds vulnerable to these infections later in life. Burcombe Hatchery, a certified organic supplier in the UK, has tested clean for major diseases in their breeding flock and strives to promote immunity with (*a*) low stocking densities; (*b*) early exposure to pathogens; and (*c*) competitive exclusion (letting the benign or beneficial microbes outcompete the pathogens).[5] These techniques promote a natural immunity that produces a robust bird and healthy environment for future birds.

Diagnosing and Treating Common Health Problems of Poultry

In my survey of Northeast organic poultry producers, the two poultry health-care books most commonly reported to be on people's shelves were *The Chicken Health Handbook* by Gail Damerow and *The Complete Herbal Handbook for Farm and Stable* by Juliette de Bairacli Levy. These two references are indispensable for the small organic poultry farmer. Damerow's book provides an excellent diagnostic tool for determining what ails the bird, and Bairacli Levy's book provides extensive herbal and homeopathic remedies. As a result of the survey and extensive research, I attempted to add to this lineup by writing *Remedies for Health Problems of the Organic Laying Flock*. This manual is a compendium of management, nutritional, herbal, and homeopathic remedies for health problems faced by laying hens. The information is gleaned from organic farmers, old poultry books, herbals, materia medicas, and the extensive Internet resources for organic health care available from the UK and India. My compendium reflects what organic farmers face when treating health problems with alternative medicine; they must piece together information, experiment, and then share their experience with other farmers. The following are ailments a producer might run into in a free-range flock of any type of poultry. The diagnostic information is from Damerow, and the remedy information is from my organic-remedies compendium.

Infectious Coryza

Also called—cold, contagious catarrh, coryza, hemophilus infection, infectious catarrh, IC, roup.

Incidence—common.

System/organ affected—respiratory.

Agent—*Haemophilus paragallinarum* bacteria.

Incubation period—one to three days.

Progression—acute and spreads rapidly, or chronic and spreads slowly.

Symptoms—in chicks at least four weeks old: depression, nasal discharge, facial swelling, one or both eyes closed, death; in growing and mature birds: watery eyes with eyelids stuck together, reddish foul-smelling discharge from nose, drop in feed and water consumption, drop in egg production, swollen face, eyes, and sinuses; sometimes diarrhea, rales, or wheezing.

Percentage affected—high.

Mortality—low, except in turkeys.

Transmission—contagious; contact with infected carrier birds and their nasal or respiratory discharges in dust, drinking water, or feed.

Prevention—avoid combining birds from different flocks and of different age groups; remove infected birds; disinfect and leave housing vacant for three weeks before bringing in new birds.

Alternative treatments—spray a mist of camphor over birds at night or place camphor oil in nostrils (do not use if treating with homeopathy); give inhalations of eucalyptus oil in boiling water in house at night; fast for one day then give laxative diet of cod-liver oil, chopped onions, and greens; give strong sage or eyebright tea; feed plenty of garlic and bran with molasses.

Homeopathic treatments—Do not use in combination with camphor treatments; homeopathic remedies for coryza are very symptom specific; the most commonly use remedies include Allium Cepa, Aresenicum Alb and Euphrashia for watery discharges; Mercurius Vivus, Hepar Sulph, Euphrasia for sticky discharge; and Nux vomica for puffy faces with a bright red color.

External Parasites

Lice

Incidence—common in laying flocks.

System/organ affected—skin and feather shafts on head, neck, vent, breast, and under wings.

Agent—most commonly head (*Cuclotogaster heterographus*) and body lice (*Menacanthus stramineus*).

Progression—lives for several months, going through its entire life cycle on the bird's body. Nits (eggs) hatch in four to seven days and are capable of laying more eggs within three weeks.

Symptoms—infested birds become irritated and cannot eat or sleep well; they may dust constantly, get restless, and injure themselves by scratching and pecking their own bodies; egg production my drop as much as 15 percent, and fertility may drop.

Percentage affected—80 to 100 percent.

Mortality—very limited except in very young birds.

Transmission—wild birds or used equipment; they spread by crawling from bird to bird or through infested feathers during molt.

Prevention—maintain robust immunity and vigorous health in the flock; provide dust baths; do not debeak since debeaked birds cannot properly groom themselves and remove lice; lime runs and houses yearly; keep wild birds out of houses; inspect birds frequently, especially around the vent for signs of lice.

Alternative treatments—provide dust baths with some or all of the following ingredients: DE (diatomaceous earth), wood ash, lime, sulfur, derris powder, quassia chip powder, or powdered charcoal; place elecampane or fleabane in nest boxes; dust individual birds with pyrethrin powder; feed garlic.

Mites

Incidence—common in litter-raised birds.

System/organ affected—skin and feather shafts; skin on shanks.

Agent—most commonly red mites (*Dermanyssus gallinae*), fowl mites (*Ornithonyssus sylviarum*), and scaly-leg mite (*Knemindocoptes mutans*).

Progression—live primarily off the bird on roosts or nest boxes; reproduce very quickly.

Symptoms—infestation causes irritation, low vitality, plumage damage, increased appetite along with low egg production, reduced fertility, retarded growth in young birds, and sometimes anemia and death.

Percentage affected—80 to 100 percent.

Mortality—limited except in very young birds.

Transmission—may live for several months off the bird; spread by contaminated equipment, shoes, or clothing; also spread by wild birds.

Prevention—maintain robust immunity and vigorous health in the flock; provide dust baths; do not debeak since debeaked birds cannot properly groom themselves and remove mites; lime runs and houses yearly; keep wild birds out of houses; inspect birds and roosts frequently, paint roosts and nest boxes with mineral oil.

Alternative treatments—provide dust baths with some or all of the following ingredients: DE, wood ash, lime, sulfur, derris powder, quassia chip powder, or powdered charcoal; place elecampane or fleabane in nest boxes; dust individual birds with pyrethrin powder; feed garlic; use *Bacillus thuringiensis*, predatory mites, or an application of silica dust; encourage *Coleoptera* beetles to live in poultry house.

Internal Parasites

Coccidiosis

Incidence—common worldwide, especially in warm humid weather.

System/organ affected—the cecum or intestinal tract.

Agent—coccidial protozoan parasites.

Incubation Period—five to six days.

Progression—usually acute, spreads rapidly, survivors recover in ten to fourteen days.

Symptoms—in chicks (cecal) or young birds: droopiness, huddling with ruffled feathers, loss of appetite, retarded growth, bloody diarrhea in early stages; in growing or semi-mature birds (intestinal): droopiness, huddling with ruffled feathers, loss of interest in water and feed, retarded growth or weight loss, watery, mucousy, or pasty, tan or blood-tinged diarrhea; sometimes emaciation and dehydration; in mature birds: thin breast, weak legs, drop in laying, sometimes diarrhea.

Percentage affected—80 to 100 percent.

Mortality—limited to high.

Inside and outside views of the author's Kingbird Farm pasture hen hoop house.

Transmission—droppings of infected birds; spread on used equipment, feed sacks, feet of humans and wild birds, etc.

Prevention—defies even the best sanitation; breed for resistance; hatch and brood chicks early in season; raise chicks on clean, dry litter to expose them gradually and let them develop resistance; avoid crowded, damp conditions.

Alternative treatments—give apple cider vinegar (1 tbs./gal.) in drinking water when chicks first arrive and when a problem is suspected; provide a variety of greens and buttermilk or raw milk to infected birds; fast birds for one day then give a senna brew with a few grains of powdered ginger; provide young birds with soil and sod.

Homeopathic treatments—if birds are listless with bloody droppings give Merc. Cor. or Ipecac; if comb is pale and there is a lack of appetite give Chellidonium or Nux vomica; if a head remedy is needed give Merc. Sol.; if birds are emaciated with diarrhea give Aconite, Merc., Podo, or Ipecac.

All Other Internal Parasites

Incidence—fairly common at low levels.

System/organ affected—all internal organs, particularly digestive tract.

Agent—round worms and flat worms (tapeworms and flukes); protozoa are also internal parasites that cause diseases such as coccidiosis (see above) and blackhead.

Progression—under good management worms and chickens become balanced in peaceful coexistence; through gradual exposure birds can develop resistance to most parasites; disease or stress usually causes an overload.

Symptoms—birds might gradually lose weight as the worms interfere with food absorption and other digestive processes; some worms, instead of invading the digestive tract, invade the respiratory system, causing breathing difficulties and gradual blockage of airways; pale head (anemia), droopiness or depression, reduced laying, foamy diarrhea, death.

Percentage affected—variable.

Mortality—limited except in very young birds.

Transmission—direct-cycle worms may be picked up on droppings or litter; indirect-cycle worms require an intermediate host such as a grasshopper or another species of worm that is consumed by the bird.

Prevention—maintain robust immunity and vigorous health in the flock; practice good sanitation; eliminate intermediate hosts (impossible on range); rotate ranges of free-range birds; avoid mixing chickens of different ages; do not raise turkeys with chickens; keep pastures clipped or grazed and limed.

Alternative treatments—give a laxative diet consisting of mash of pumpkin seeds and milk, and after a twelve-hour fast, follow with a warm mash of bran, middlings, and milk; feed finely chopped onions, garlic, carrot, parsley, pumpkin seeds, carrots seeds, or fennel seeds; try the following herbs: elder leaves, wormwood, wormseed, cotton-lavender, rue, hyssop, cayenne, senna, male fern, castor oil, wild ginger, snakeroot, goosefoot, conifer needles, fennel seeds, or pyrethrum all preceded by a fast and followed by a laxative period.

Homeopathic treatments—if worms in the nostrils or ears give Cina; if affected birds lose weight, ruffled feathers, rapid breathing give Aconite, Santonite, or Tucrum merver; if a head remedy is needed give China.

Pasted Vent

Also called—cloacitis, vent gleet.

Incidence—common in chicks.

System/organ affected—vent, cloaca.

Agent—unknown. May be due to improper consistency of droppings caused by rations or chilling; may also be due to unsanitary practices at hatchery.

Progression—chronic.

Symptoms—in chicks up to ten days old: droopiness, droppings sticking to vent; in hens: offensive odor from droppings sticking to vent feathers.

Percentage affected—usually limited.

Mortality—possible if vent gets sealed shut.

Transmission—does not spread from bird to bird.

Prevention—keep chicks warm; do not hatch chicks from affected hens.

Alternative treatments—carefully remove dropping from vent with warm water; give apple cider vinegar *or* bicarbonate of soda in drinking water (1 tsp./quart); apply weak iodine solution to vent; give a laxative diet followed by garlic.

Homeopathic treatments—give Ipecauncha with Camomilla or Carbo veg.; if discharge is bloody and white give Merc. sub Cor. or Ars. Iod.

Picking

Also called—cannibalism.

Incidence—common in high-strung breeds or those in close confinement.

System/organ affected—toes, head feathers, back feathers, vent.

Agent—breeding; crowding; bright lights in brooder; boredom or lack of exercise; feed and water troughs too few or too close together; feed too high in calories and too low in fiber; external parasites; injury or bleeding.

Progression—some picking like vent picking can escalate rapidly once blood is drawn.

Symptoms—dull broken feathers, bare patches; bloody raw vent.

Percentage affected—limited to flock-wide.

Mortality—vent picking may lead to death.

Transmission—a learned behavior that can spread rapidly through a young flock.

Prevention—reduce light intensity; spend time with birds, handling and walking among them; observe birds to identify and isolate bullies (especially turkey poults); use more docile breeds; avoid overcrowding and stress; provide birds with forage or range; in winter provide birds with whole grains, alfalfa hay, or root vegetables; ensure a balanced diet.

Alternative treatments—apply bitter apple or Listerine to picked area; give 1 tsp. salt/gal drinking water in the morning, then again in the afternoon and repeat three days later; nail a piece of raw salt pork or beef suet to the wall for pecking; hang bunches of nettles, comfrey, or heads of cabbage; for hens, reduce egg size by reducing levels of linoleic acid and high energy when hens do not need it; increase fiber in diet.

Homeopathic treatments—give Helleborus niger.

Table 9. Toxic Plants

Common name	Botanical name	Toxic parts
Blue-green algae		all
Bryony	*Bryonia cretica*	
Buttercup	*Ranunculus* spp	
Castor bean	*Ricinus communis*	bean
Clematis	*Clematis* spp	
Corn cockle	*Agrostemma githago*	seed
Daubentonia	*Daubentonia longifolia*	seed
Deadly nightshade	*Atropa belladonna*	
Death camas	*Zygadenus* spp	leaf, stem, root
Glottidium	*Glottidium vesicarium*	seed
Hemlock	*Conium maculatum*	
Henbane	*Hyoscyamus niger*	all
Horseradish	*Armoracia rusticana*	
Iris	*Iris* spp	
Laburnum	*Laburnum anagyroides*	
Milkweed	*Asclepias* spp	leaves
Monkshood	*Aconitum napellus*	
Nightshade	*Solanum nigrum*	immature berries
Oleander	*Nerium oleander*	all parts
Pokeberry	*Phytolacca americana*	berries
Potato	*Solanum tuberosum*	green tubers, raw peels, and sprouts
Privet	*Ligustrum vulgare*	
Ragwort	*Senecio* spp	
Rapeseed		
Rhododendron	*Rhododendron* spp	
Rhubarb	*Rheum palmatum*	leaves, flowers
St. John's wort	*Hypericum perforatum*	leaves, flowers
Sweet pea		
Vetch	*Vicia* spp	pea
Yew	*Taxus* spp	all parts

Sources: Gail Damerow, *The Chicken Health Handbook;* Victoria Roberts, *Disease of Free-Range Poultry* (Suffolk, UK: Whittet Books Ltd., 2000); Katie Thear, *Free-Range Poultry* (Ipswich, UK: Farming Press, 1999).

Poisoning

Birds in free-range systems may be more susceptible to natural poisoning than confined birds. While organic birds are not usually exposed to chemical poisons like herbicides, rodenticides, parisitides, pressure-treated lumber, or strong disinfectants like commercial confinement birds, they may face other dangers. When on range, birds may encounter a number of poisonous plants in paddocks, along fences, or in hedgerows. Pastures should be scouted yearly to identify and remove toxic plants. Maintaining abundant high-quality forage for the birds to graze will also help prevent the consumption of undesirable plants. See table 9 for a list of plants toxic to poultry.

Mycotoxins could also be a problem in old or spoiled feed (see table 10). Many free-range birds are fed directly on pasture where feed can spill and easily become moldy. Typically birds with fresh feed available to them will not bother with spoiled feed, but if or when their feed is restricted, as in the case of rapidly growing broilers, they may be willing to ingest moldy feed. Feed quality should also be carefully monitored in storage. Nux vomica is a common homeopathic remedy for poisoning, but poison-specific remedies (isopathy) like Belladonna for nightshade poisoning and Hypericum for St. John's wort poisoning are helpful when the source of the poisoning is known.

Table 10. Fungal Poisoning.

Disease	Caused by	Grain source
Aflatoxicosis	*Aspergillus flavus* and other fungi	all grains
Ergotism	*Claviceps purpurea*	wheat, rye, cereal grains
Fusariotoxicosis	*Fusarium sporotichioides*	corn, wheat, barley, millet
Ochratoxicosis	*Aspergillus ochraceous* and other fungi	barley, corn, sorghum, wheat

Source: Damerow, *The Chicken Health Handbook.*

THE AUTHOR'S EXPERIENCE: Health Problems of the Older Hen[6]

There are flock-replacement issues faced by commercial flock owners, but what about those of you who want to keep your small flock for the extent of their lives? Backyard flock owners often keep hens well into their geriatric years despite the drop in production. These poultry keepers find they face many different health issues as the hens age.

While production efficiency goes down precipitously as hens age, they can continue to be healthy and productive for many years. These older hens can, however, be susceptible to health problems that do not usually crop up in short-lived commercial flocks at the peak of their vitality and productivity. The following health issues are problems I have encountered mainly in older hens, particularly those that have lived in the same housing for many years. Young flocks that are rotated out of seasonal housing and spend their production year outside on rotated pasture experience these problems more rarely.

Bumblefoot

Bumblefoot is lameness caused by foot lesions infected with the *Staphylococcus aureus* bacteria. Staph can infect hens' feet when rough roosts, stony ground, or chicken wire injures them. I have also seen staph develop in feet and shanks injured by a tangle of the incredibly strong, fine string used to sew closed the tops of feed bags. Injured feet exposed to wet dirty litter or muddy yards that have held chickens for many years are especially susceptible. The first step to prevention is to remove any surfaces or objects that might injure chicken feet. Ensure that perches are well designed; circular hardwood perches with a flattened upper and lower surface are ideal. Keep litter dry and friable but not too coarse. Eliminate sharp objects such as loose chicken wire, exposed nails, broken perches, bent nest box edges, rusty metal feeders, or old equipment in the chicken yard.

If a hen does become lame there are effective remedies. Wash the affected foot thoroughly, open the abscess, squeeze out the infected core, clean with hydrogen peroxide or calendula tincture and wrap the foot with gauze or vet wrap. If you want to allow the abscess to open naturally soak the foot in Epsom salts or apply a warm linseed-meal poultice wrapped with gauze. If there is no abscess, but instead dry, rough cracks, apply tea tree oil to the cracks and keep the bird on a soft clean surface. Common homeopathic treatments are Hepar sulph or Silicea orally in combination with calendula cream on the affected foot. Under any treatment, hens should be kept clean and quiet with good nutrition and ample water. Check gauze or other wrapping frequently so it doesn't become loose or tangled.

Scaly Leg Mites

Scaly leg mites are tiny external parasites called *Knemidocoptes mutans* that infect only the shanks and feet of older hens. These mites feed under the hen's shank and feet scales raising the scales by generating debris that accumulates beneath them. The feet and shanks become thickened, crusted, and very irritated. These mites spread by traveling from bird to bird on roosts or nest boxes and are very difficult to control.

Topical treatments for leg mites include scrubbing feet with soapy water and ammonia then rubbing in a solution of garlic, cayenne, and vinegar. You can also try smothering the mites by applying a coat of mineral oil, lavender oil, or Vaseline to the feet and shanks every day. Internally it is helpful to boost the bird's defenses by giving a free-choice mineral mix, kelp, and fresh greens. Sulphur pellets in the drinking water every day is a useful homeopathic remedy. Roosts and nest boxes should also be treated. Apply pyrethrin dust to the cracks of roosts and nest boxes and use cedar chips in nest boxes during the treatment. This is a very difficult parasite to eliminate.

Body Lice

Chicken body lice (*Menacanthus stramineus*) are the most common lice to infect older hens. The lice are tiny and straw-colored and can be seen moving quickly away when the feathers around the vent are parted. This louse leaves clusters of white eggs (nits) attached to the feather base primarily around the vent and under the wings, but also on the breast and head. These parasites are found in most flocks, but typically a healthy, robust hen living on rotated pasture will not succumb to them. Any flock, however, can have a flare-up while confined to winter housing, particularly when their health is compromised by a cold or other infection. Severe infestations can seriously reduce production since the hens are constantly irritated and cannot sleep.

Aside from ensuring good overall health and immunity, it is important to allow hens to properly groom themselves. This includes not debeaking them and providing them with effective dust bathing areas. Debeaked birds cannot preen properly and remove the nits from their feathers. Birds not provided with a good dust bath will often dust in the litter creating fecal dust and potential respiratory problems. Dust baths can also cause respiratory problems so it is important that they not create too much airborne particle matter that could be irritating to the hens. Many different ingredients can be included in a dust bath, but some are more effective than others. The benefit of the bath must outweigh the potential harm to the hen's lungs (and yours). I have found a good bath mix to be wood ash, agricultural lime, diatomaceous earth (DE), and cedar chips. I incorporate this mix into the driest section of the bedding where the hens typically dust anyway. When the dusty ingredients are mixed with cedar shavings and bedding they create less airborne dust.

Should a severe outbreak occur, birds can be dusted individually with pyrethrum and DE. It is very important to protect your lungs and the bird's lungs when doing this. I wear a good dust mask and dust the birds with the "shake and bake" method. This involves putting the bird in a small feed bag containing and equal mix of DE and pyrethrum powder with the bird's head protruding from the bag opening. I then hold the bag closed and shake gently making sure I get the mixture up into the downy feathers around the vent. This technique is best undertaken at night to keep the birds calm. You want to place the birds gently back on their roosts so the mix stays in their feathers all night. Pyrethrum has a short effective period, especially once exposed to light, and if this method is used during the day, or the birds shake it all out of their feathers, it probably will not be very effective. Other treatments for lice include applying lime to runs or bedding; painting roosts and nest boxes with mineral oil; adding cedar, elecampane, fleabane, or wormwood to nest boxes; or feeding garlic (this will affect egg flavor).

The foundation of any health-care system for your hens should be good nutrition and healthy living conditions for strong natural immunity. With these key advantages hens should lead long, productive lives. However, as with any animal, aging will have an effect on a hen's ability to maintain production and health. Giving her a little extra support and carefully observing her needs and behavior will go a long way to extending her productive life.

---

six

Specific Management Challenges

Organic Broiler Management Issues

Although organic poultry-management principles may be used across the board for many different poultry enterprises, some specific problems or issues face growers of pastured organic broilers. For example, birds have a slower growth rate in free-range systems, and they may consume extra feed to stay warm, cool down, walk to feeders, or escape predators; thus farmers sometimes see no reduction in feed consumption with birds out on pasture.

On the other hand, the primary advantages of growing organic broilers in a free-range system include the fact that the birds' slower growth rate prevents health problems such as ascites (water belly)[1];and free-range exercise and slow growth reduce leg problems.

So, although free-range birds tend to grow slower and potentially consume more feed, they are generally healthier and more properly developed than confined birds. This can be a balancing act for the organic producer who is trying to grow a healthy vigorous bird on as little expensive organic feed as is appropriate. Most producers recommend careful monitoring of feed intake to ensure that the birds are not wasting it on the ground or converting it to fat. This can be accomplished by designing feeders properly and filling them appropriately and working with a poultry nutritionist to guarantee the proper feed ration.

Barred silver cockerel meat breed.

THE AUTHOR'S EXPERIENCE:
Organic Broiler Production[2]

Poultry are small, low-investment livestock with steady consumer demand. Free-range poultry seemed only natural and our neglected pastures at first craved nitrogen and a good scratch. Wanting to move beyond the classic chicken yard model, we were drawn to Joel Salatin's pastured poultry system. It allowed a little "free" range and yet evenly distributed that precious nitrogen across our fields. However, within a year we had a row of abandoned pasture pens along our fence. We weren't satisfied with the space they afforded the birds or the constant pen moving. Despite our fellow producers' declaration that Cornish Rock Cross broilers would not choose to run, jump, flap, and forage if they had the option we went to a nearly complete free-range system.

We currently use a 12 × 24 ft. hoop house within two 150 ft. electric nets to raise four batches of 200 broilers. The Cornish Rock Cross broiler chicks (straight run) typically come from Moyers Hatchery. We start batches of 200 in a 12 × 12 ft. brooder. Depending on weather conditions, the chicks are allowed out onto a 10 × 12 ft. net-covered "poultry patio" at seven days. This begins a week or two of "hardening off" when they can experience the perils of weather and learn to forage while still having the safety of the brooder. We believe this simple outdoor access has greatly reduced early mortality and better prepared the chicks for a free-range life.

By three weeks or less the chicks have clearly outgrown the space and are ready to move out. They have had free access to a certified organic 20 percent ration and chick grit. They have learned to drink from a hanging bell drinker and forage through the hay, weeds, and vegetables we have provided on the patio. They know to seek shelter when it rains and not pile up in the corners when the heat lamps are gone. We load them into crates on a nice day and drive them up to their first paddock, which is approximately 100 × 50 ft. and enclosed by two Premier electric nets. The hoop house there easily offers enough shelter for 200 mature broilers. Metal trough feeders are placed in front of the house to encourage the birds to use the paddock. Each day the feeders are moved down the paddock one length. This distributes the manure more evenly and helps the birds better utilize the space. Contrary to the warnings, they do not just belly up to the bar; they run, they forage, they dust, and sometimes even fly. They have had no leg problems and we do not run over them with the pens. They thrive in this environment.

At approximately eight weeks we process the broilers on-farm. We have a very simple setup in an outdoor shed. In the dark early morning the birds are loaded into crates and driven down to the processing shed. We butcher each bird by hand to ensure humane handling and high quality from beginning to end. We want to see every step of the process, just as we did raising them. Our scalder and plucker were self-constructed and cost far less than commercial equipment. Our countertops are stainless steel and kept scrupulously clean. After eviscerating and cleaning, the birds are chilled in ice before being bagged. They are weighed and labeled in our sales room. Preorder customers pick them up fresh from our commercial cooler. Those birds not

Kingbird Farm pastured broiler hoop house.

preordered are frozen after twenty-four hours and sold at the farmers' market.

The key to our operation has been quality: a rare and desirable product raised in a manner that pleases the customer and involves the whole family. We take pride in our birds and the lengths we go to ensure their quality of life. In addition, the birds benefit our farm in many other ways. They have improved our pastures with their manure, dethatching, and insect removal. They have added to our farm diversity without taking over, and they have provided a reliable income. While still only one part of our farming system, the broilers have proved worthwhile and enjoyable.

Organic Layer Management Issues

Producers of organic eggs face additional challenges. Not only are they expected to get good production on organic feed in a free-range system, consumers also want the product to look a certain way: brown-shelled eggs with firm whites and that brilliant orange yolk, year round. The farmer must work diligently to ensure this quality. Organic egg farmers cannot afford to dump overproduction or low-quality eggs onto the conventional market; they have too much invested in the organic hen. Eggs must be premium and command a premium price. Organic free-range growers and, to a certain extent, non-pastured confinement organic growers tend to face the following issues:

- Slightly lower production on pasture.
- Potentially dirtier eggs on range due to weather, pasture quality, and nest boxes.
- White-egg syndrome, which causes brown-egg layers to produce very pale-shelled eggs when exposed to sunlight (see "Brown Shells" section below).
- Fading yolk color with birds off pasture in the winter.

Production and culling

Although free-range egg growers tend to see a slightly lower production rate than commercial confinement operations, they can also have fewer inputs and lower overhead. In addition, the value of the eggs is so much higher it tends to make up for lower production. Still, an organic producer cannot afford to feed and manage a hen that is not producing. Selective culling throughout the season helps maintain production but requires the producer to know her flock through time spent observing the birds. Some smaller producers tend to keep hens for years without monitoring their production and this can be economically unviable.

Aside from replacing the entire flock after the first or sometimes second cycle, producers should also cull during production. No matter what production system is being used, it is advisable to cull during the night when birds are least active or prone to being disturbed. Handle birds gently, starting on hens with obviously shrunken or pale combs. Use a headlamp to keep hands free to check birds for:

- Comb color: Pale, dry, shrunken combs are bad. Birds in production have plump, red, waxy combs (sometimes drier in winter).
- Leg color: Yellow is not good. Birds in production have pale legs.
- Weight: The ones that startle you with their lightness or heaviness should be culled.
- Distance between the pubic bones: Three fingers' width is good.

THE AUTHOR'S EXPERIENCE:
Thoughts on Culling the Flock[3]

As fall draws near, when the weather grows cold and the grass stops growing, it's time to think about culling the flock, to decide who will make the transition to winter housing to lay another six months. The eighteen-month-old hens have been laying now for twelve months. They are molting and not likely to be as productive next year. The six-month-old hens that began laying this spring will continue through the winter and be culled next spring, after twelve months of production. Thus, every six months part of the flock is culled and a new set of pullets comes in to take its place. Consequently, in fall I remove approximately 150 eighteen-month-old hens and move the remaining 150 twelve-month-old hens to winter housing where 150 six-month-old pullets are waiting and just starting to lay. The following spring we repeat the process sending them out to summer housing with a new batch of newly laying pullets we have raised up over the winter. This system gives us consistent production from a vital flock of 300 young, healthy layers, half of which are newly laying and half of which are part way through production. It also requires us to cull 150 hens every six months.

Culling is no one's favorite task. However, after their second season the hens decline in production enough that they cease even to pay for their feed—fine in a backyard flock, but threatening to the viability of our organic egg business and the quality of management we can give to the overall flock. The hens must support their corner of the farm.

There are several options for a hen retired from commercial production. We use all of the following options when culling and hope to make the best decisions both morally and financially:

1. Sale to a person with a backyard flock.
2. Processing for sale as stew hens.
3. Culling and composting for unthrifty or injured birds.

We choose the option based on the breed, health, and production condition of the hen. We initially try to sell the healthy retired hens to people with small backyard flocks who don't require the level of production we do. Eighteen-month-old hens are plenty productive for the small flock owner as long as the hens are healthy and well cared for. We base our price for these hens on what we would get for that bird if we processed it as a stew hen. Depending on breed, a dressed stew hen weighs 2½ to 3 lbs.—worth $5.63 in the freezer, but we have to process, package, freeze, and sell it. If someone buys that bird live off the farm, we have no additional labor. Based on those numbers we typically sell retired certified organic laying hens for $4 picked up at the farm.

"Least Favorite" Choice

After selling as many as possible live, we look at the health and body condition of the remaining hens to make our final cull. Hens that are obviously unthrifty or injured are immediately culled and composted for their own sake and the well-being of the flock. This is done throughout the season as well as during the major culling periods. Some breeds are very slight and never have much meat on them; these are rarely worth processing.

Culling and composting is our least favorite method. It only requires us to kill and properly compost the birds but sometimes feels wanton. Somehow, the killing and dressing of a retired bird for human nourishment seems a fair choice, but composting can feel cruel and pointless. Of course it's not cruel and pointless, especially in the case of a sick, suffering bird. The carefully tended compost of chickens, manure, and straw is turned and allowed to heat for a year to produce a vital, nitrogen-rich food for our crop soil. The composted birds return to the earth and help maintain the vitality of our farm.

The hens we choose to process for eating must be healthy, robust, and not molting. For the effort we want a fairly large (2½–4 lb.), plump bird with good fat coverage (for that amazing broth) and no pinfeathers. The lack of pinfeathers is particularly important if you're attempting to clean colored birds like Rhode Island Reds, Black Sex-Links, or Barred Rocks. White birds like Delawares, Leghorns, and White Rocks are usually much easier to clean regardless of molt. Older birds can be more effort to pluck, eviscerate, and cut up. Tendons, ligaments, and skin are much harder to cut and thus more effort to process. Keep this in mind if you are planning to process your own stew hens; they take more time than broilers and typically bring in less money. That being said, there is a market for organic stew hens.

They are virtually unavailable in grocery stores and make the most marvelous chicken soup or stock. Their flavor and richness far surpass a broiler chicken, but they must be stewed long and slowly for tender meat. Remember, they are usually at least sixteen months older than a broiler and have had a lot of exercise (at least free-range hens have). They have also had a lifetime of foraging for seeds, insects, and greens and a balanced diet of certified organic grains. This diet produces rich golden fat for broth and delicious meat.

With these thoughts about culling in mind, carefully consider the breed, age makeup, and management of your next laying flock. We are conscious of trying to select a breed with the following criteria:

1. Good production in a free-range system.
2. Docile, levelheaded, with social skills.
3. Medium to heavy with good feathering for winter.
4. Robust, meaty, and worth processing for stew.
5. Not so fat and heavy that feed converts into fat and not eggs.
6. Good foraging habits.
7. Light-colored or white feathers that are easy to remove.

Finding a breed with all the above criteria is rare, and we are always trying new breeds to maximize the use of the bird overall. Eventually, all laying hens must move on, and it is up to the farmer to select a good breed both for laying eggs and what lies beyond.

Egg Cleaning

The main focus of any egg-cleaning method should initially be clean nest boxes. It is much easier to produce clean eggs than to clean dirty eggs. Eggs are porous, and improper washing can actually introduce contaminants. Consumers perceive organic eggs as safer and cleaner than conventional eggs, but this cannot be true if they come in from the field caked with mud and manure—especially if improperly washed.

Most small producers clean eggs by hand, while larger operations find it more economical to use egg washing machines. When cleaning by hand it is important to follow some basic rules to maintain egg quality and safety.[4]

1. Wash eggs as soon after collection as possible.
2. Dry clean mildly dirty eggs with sandpaper, steel wool, or loofah.
3. If eggs are very dirty consider using them for another purpose.
4. Wash and rinse water should be 90°F–120°F—at least 20 degrees warmer than the eggs. Between 110°F and 120°F is best since this temperature causes the egg to expand and prevent entry of any microbe-contaminated water through the shell pores.
5. Wash water *can* include an unscented, non-foaming detergent (check with the organic standards before adding anything).
6. If an immersion-wash machine is used with baskets or flats, change water often.
7. If eggs are hand-cleaned individually, use paper towels or bleach-washed cloths that are changed frequently.
8. Presoak only briefly before washing.
9. Prepare a very dilute solution of bleach for rinsing and sanitizing.
10. Use chlorine test strips to test solution for 50 to 200 ppm of free chlorine.[5]
11. Allow eggs to air-dry before packing.

Brown Shells

In the United States, the public perception is that an organic egg is a brown egg. Thus it behooves most farmers who produce shell eggs to maintain good shell color. This can initially be done of course by selecting varieties of hens that are known to produce dark-shelled eggs. However, even brown-egg layers can lay pale, almost white eggs due to a variety of factors. Birds tend to lay paler eggs as they age, and many small producers keep their birds for several cycles. Diseases such as infectious bronchitis,

Newcastle disease, avian pneumovirus, egg-drop syndrome, heavy parasite infestation, or stress can also cause a loss of shell color.

By far the most interesting cause for the loss of eggshell color is known as white-egg syndrome. In the UK this problem has been discussed for years among free range producers who have found that hens exposed to direct sunlight lay pale-shelled eggs.[6] These hens are commonly high producers that also have poor feathering on their backs due to picking or nest-box breakage. The producers have found that the problem can be quickly remedied by moving the birds indoors. No definitive cause for the loss of pigment only during bright sunny weather has been found. It has been suggested that the stress combination of bare back exposed to direct sun is enough to compel the hens to lay eggs before the full amount of pigmentation has been laid down on the cuticle. The present information on this syndrome is anecdotal, but deserves further observation and research.[7]

Yolk Color

Despite the fact that yolk quality is more important than yolk color, this is the factor by which consumers often gauge the quality of a free-range organic egg after cracking it. Yolk color is formed by carotenoid pigments that occur in green plant material and corn. The most effective carotenoids are lutein and zeaxanthen that are commonly found in dark leafy vegetables.[8] While a free-range bird on pasture will most likely ingest just enough plant material to color the yolks, it is not wise to force a lot of greens on hens. A high intake of green plant material may give a slight khaki color to the yolk and possibly even greenish-yellow tint to the albumin. Various plants such as shepherd's purse give a slight olive color to the yolk.[9]

Laying hens on pasture or other managed forage will most likely maintain a good yolk color during the growing season. To continue this yolk quality into the winter, many producers offer supplements such as green-leafy alfalfa, sprouted oats, dried comfrey, or free-choice whole corn. Many semi-confinement organic operations, such as The Country Hen in Hubbardston, Massachusetts, include ingredients like marigold flowers in their feed ration.[10] Marigolds contain lutein and not only enhance egg-yolk color but offer some health benefits for the consumer as well.

Organic Turkey Management Issues

Organic turkey raising can be successful when done right. Pennsylvania alone produced 70,000 certified organic turkeys in 2001.[11] Turkeys thrive in low-stress, free-range environments when they can be properly protected from predators and inclement weather. Even the standard broad-breasted white turkeys forage aggressively and range efficiently. The key problem for small farmers is growing turkeys and chickens on the same land. Most small organic farms are quite diverse and raise a variety of poultry. In my survey of small Northeast growers one-third of the egg producers also kept turkeys. This can be risky since turkeys are uniquely susceptible to numerous diseases carried by chickens, waterfowl, and wild birds. Small farms might get away with co-raising turkeys with other poultry for several years, but eventually something will strike. Organically raised turkeys are too high in value to risk them in this way.

Small producers can successfully grow turkeys if they take basic steps to prevent contamination from other poultry.

- Turkeys should have separate facilities: brooder, housing, range, or pasture.
- Turkeys should have separate equipment: feeders, drinkers, brooder lamps, buckets, hoses, fences, feed bins, etc.
- Turkey chores should be done first thing in the morning before the rest of the farm chores or by a separate person.
- A footbath outside brooders and housing or a separate pair of boots should be used for turkey chores.
- Minimize spilled or excess feed on range to avoid attracting wild birds.

THE AUTHOR'S EXPERIENCE:

Organic Turkeys–Trial by Fire[12]

A Little History

The several hundred turkeys we've raised on Kingbird Farm have collectively caused more strife and trepidation than the thousands of broilers that have passed through. Each batch came with its own challenges and sometimes even rewards.

The first couple of years were relatively uneventful. We raised small batches in pasture pens without preselling them and got our feet wet. In year three we were still brooding and raising the turkeys right alongside the chickens. This was the first year we began to see problems, though they had nothing to do with the proximity of chickens. The feed ration was probably imbalanced, and we started to see slipped tendons on the poults at two weeks of age. We quickly supplemented with hard-boiled eggs and fresh greens, and most of the poults recovered.

In year four the poults did quite well as youngsters and we put a large, healthy flock out to pasture. As November rolled around we clued in that the birds were not growing very large. They were healthy white turkeys that were flying, roosting, and foraging very well, but not growing. They were long and lean and quick, like wild turkeys. Well, we had deadlines and the birds were processed the Sunday before Thanksgiving regardless. These vibrantly active birds dressed out to an average of 10 lbs. at sixteen weeks. Needless to say, those hoping for a 20 lb. turkey did not get it. We ended up giving several people two turkeys.

In year five, in an effort to ensure big turkeys, we started the poults a full month earlier. This was also the first year we provided an entirely separate brooder and pasture for the turkeys. The batch thrived and thrived and thrived. They aggressively consumed all the pasture we had for them and bulked up on the feed as well. By the last month we were feeding them practically nothing but whole oats and grass in an effort to slow their growth. These behemoths dressed out to an average weight of 25 lbs. in nineteen weeks. This was not the happy medium we were looking for. However, nobody could complain about small turkeys.

Catching Cold

The next year, year six, we thought we held all the cards. We received our poults on July 3, we had a good handle on the feed ration, the facilities were all separate from the chickens, and the weather was ideal. But we did not expect the sniffle. One of our barnyard chickens probably had the sniffles. This chicken probably sneezed near the turkey brooder. Since the brooder is separate, but not on the opposite end of the farm, the newly arrived, immunodeficient turkey poults caught that sniffle. They took the sniffle very seriously and started dropping like flies. Their heads swelled up, their eyes watered and they sneezed constantly. I quickly began treating with fresh greens, sod, and homeopathy. I wasn't immediately concerned with an impending disaster. I always only presell half the turkeys, and well over half were still relatively healthy. I banked on recovery and proceeded with caution.

As the days progressed, so did the loss of poults. We contacted Cornell and had a necropsy done on two birds. The diagnosis was quick and straightforward: sinusitis. But turkeys can be wimps, and this was a perfect example of how fast a simple sinus infection can sweep through a flock. So to cut my losses I moved the flock to the other end of the farm. I set up a horse trailer as an impromptu brooder and began treating the sick turkeys with a number of different herbal and homeopathic remedies. The doctor at Cornell did not foresee that the use of antibiotics would dramatically help the already sick birds, so I struck out on my own.

That year we completely cancelled turkey sales, holding the few surviving birds for friends and family.

What We Have Learned

Track your poult order. One of the first challenges we faced was getting turkey poults when we wanted them. Many hatcheries stop hatching in June, and we don't want to start them until the first week of July. If you can process well before Thanksgiving and freeze the birds, you can be more flexible. We process the Sunday before Thanksgiving, not before and not after. If you can order the poults for the date you want them, make sure they actually come. More than once we have called to confirm an order only to find there is a shortage and the hatchery wasn't planning to send us any turkeys at all. This leaves us scram-

bling to find another hatchery on very short notice. Once we had to supplement our order from another hatchery that also ran short of poults and didn't send them for another two weeks. When the poults finally arrived they were debeaked and toe-clipped! Fortunately the poor little guys managed to forage on pasture and actually caught up to the older birds we had received with our original order.

Develop an excellent feed ration. Young turkeys cannot be started simply on broiler feed with much success. Most broiler feed is too low in protein for poults. We initially used only broiler feed, supplementing with hard-boiled eggs and beef liver, but this was very inconsistent and had varied results. We now provide them with a 23 percent turkey starter while in the brooder. Once they are fully feathered and out on pasture we switch to the 21 percent broiler feed. Depending on growth rate, we may later cut this feed with whole oats and corn as a finisher. It is also important to provide turkeys with age-appropriate grit and free-choice sod/greens when in the brooder. Turkeys are nutritionally sensitive when young, and it is critical to get them through this period with a well-balanced, complete ration.

Establish separate facilities. For years, seasoned turkey growers told us some day we would have our turkeys wiped out by a chicken disease. They were right. It is difficult to raise turkeys and chickens on the same farm. We were just lucky for five years. It finally hit us hard.

1. Build a completely separate brooder space for the turkeys that is as far from the barn as possible and still has access to water and power (approximately 300 feet from the nearest chickens).
2. Establish and maintain completely separate range housing and pasture (we built a 12 × 12 ft. roost shed with two acres of rotating pasture around it).
3. Maintain separate feed storage (the turkey feed goes directly from the truck to the turkey house).

The turkey challenge. Small producers can grow turkeys if they carefully prevent contamination from other poultry; use a separate set of equipment and a separate tending regimen.

4. Assign a single chore person who will tend the turkeys first thing in the morning before going to the barn.
5. Maintain a boot bath outside the turkey housing.
6. Do everything possible to boost health and immunity.

You Can Enjoy Them

Despite the trials we have endured raising turkeys, we continue to enjoy raising them and providing them for our customers. Turkeys are delightful, friendly birds that thrive on good forage and open air. We believe the key to their success is vigorously supporting their health from day one and keeping them separate from other poultry.

Waterfowl for Organic Production

Waterfowl are generally well suited to organic free-range systems. Ducklings and goslings are fairly trouble free to brood and have always been raised without coccidiostats and antibiotics, both of which are toxic to waterfowl. Ducks and geese range well, herd easily, and once fully feathered, can handle harsh weather. Ducks forage well, which reduces feed expense and produces flavorful meat and eggs. Geese are true grazers and can be raised entirely on forage of good quality. Organic ducks and geese are still a relatively rare commodity and command high prices. The limiting factor for many small producers who do their own processing is the extra plucking and waxing needed to clean waterfowl.

Waterfowl can be used in many innovative ways on the organic farm. Weeder geese were used on a large scale before the widespread use of herbicides. Muscovy ducks are used for fly control on dairy and other livestock farms. While the use of ducks for fly control is probably completely allowable under current organic standards, weeder geese could be problematic. Geese have been successfully used to weed strawberries, potatoes, onions, garlic, cane berries, cotton, mint, and other herbs. But under current organic certification rules the geese would probably need to be removed from the crop 120 days before harvest of berries, cotton, or mint and 90 days prior to the harvest of potatoes, onions, or garlic. After their removal their droppings would need to be incorporated into the soil through cultivation. This removal period gives ample time for a new flush of weeds to emerge, thus defeating much of the purpose in annual crops. In long-term perennial crops, geese could possibly be used in the off season to control perennial weeds provided the weeds are young and tender enough for the geese to desire and utilize for nutrition. The innovations of small organic growers will continue to explore and challenge the rule in circumstances like this.

Geese, true grazers, can make excellent use of pasture.

THE AUTHOR'S EXPERIENCE:
The Pastured Duck

Pasture-raised duck has been a very successful part of poultry farming on Kingbird. We started when customers began to request duck for special occasions. Initially my husband was concerned about the economic viability of the project, since ducks are more difficult to process. However, we found processing to be the only difficulty in raising duck in our system.

Brooding

We chose Pekin ducks to begin with for their uniform growth and white feathers. An unused greenhouse became their brooder, with ample sunshine and warmth for spring brooding. When temperatures increased beyond the comfort of the ducklings we covered the greenhouse with a shade cloth and provided fans. We initially sectioned off the greenhouse to start the ducklings in one corner and then gradually gave them more space as they grew. Unlike chicks, they could not fly over the low barriers we used to divide the space. For brooding we used the basic equipment used for starting any poultry: infrared heat lamps, 1-gallon chick drinkers, dry softwood shavings, and paper egg flats as feeders. The only adjustment we made to this setup was to place the drinker in a garbage can lid or other low-sided lid to prevent the bedding from getting wet. Ducklings tend to splash and sputter large sums of water everywhere. They enjoy running back and forth between the drinker and the feed, alternating mouthfuls, dirtying the water and moistening the feed. All of this duckling activity can lead to moldy feed, smelly water, and soggy bedding if it's not managed. We change feed and water more often with ducklings, but not excessively.

Outdoor Access

As the ducklings begin to feather out and the weather turns warm, we begin to allow them outdoors during the day and herd them back in at night. Letting them out gives them access to forage (which they relish) and allows us to feed and water them outdoors and avoid litter management issues. We do not allow them swimming water (kiddie pool) until they are fully feathered—then we observe to make certain they can easily get out.

We have managed ducks as both free-range on the pastures and confined within netted paddocks. Both methods have their pros and cons. Since we have little predator pressure we initially released them onto the pastures during the day and herded them home in the evening. This allowed them access to several acres of prime pasture—and they used it. The method ceased being workable when the entire flock discovered their passion for ripe tomatoes! This led us to the current method: rotating paddocks fenced with electric netting. Ducks do well with netting considering they are so well insulated. The first few days they may go through it, but after a while they respect the nets and behave themselves. They will, however, go right through it or even pile over en masse if pressured by dogs or aggressive children. We find that the ducks graze well on fresh growth if moved fairly often. In addition to their forage they receive a full ration (broiler/grower feed) that we feed twice a day with ample water and expect them to finish between meals. Limiting their feed in this way promotes grazing and reduces extra layers of fat later in growth.

Processing

The tricky part of duck raising is indeed the processing. They must be processed at exactly the right time or feathers and down will be in pin and impossible to remove. We typically process our ducks at eight weeks and expect a 3½ to 4½ lb. carcass. On processing day we herd the ducks down to the processing shed and pen them up with hog panels. We also hang tarps around the hog panels to keep the ducks calm and be sure they have ample water while they wait. We follow the basic poultry processing routine with a kill cone, scalder, and barrel plucker. The only additional step is waxing and stripping to deal with the down and pinfeathers. This involves dipping the carcass in melted duck wax (see appendix 1), then dunking it in ice water, stripping off the hardened wax, and repeating the process. This extra step just about doubles our processing time, which needs to be made up in the price of the bird. Since we find the ducks so easy to manage and we typically have zero mortality, the only extra labor is in processing. We package and sell the ducks as whole birds and currently charge $4.50/lb. We are pleased with the profit margin on duck, although it is probably a limited market and we only offer about 150 each year.

---

seven

Organic Slaughter and Processing

The Basics

Certified organic meat, including poultry, cannot be marketed as certified unless it has been slaughtered and processed in a certified organic plant. Most small producers cannot justify economically the expense of such facilities. These producers, however, have another option. There is a federal 20,000 bird per year freedom-of-inspection law (The Poultry Products Inspection Act, section 15) for small producers to process on-farm. This law applies only to poultry and restricts the producer to in-state shipping. Often tighter state and local laws override this federal law. Producers must research what is required in their location. In addition, your organic certifier may require documentation of processing methods and/or organic handling certification. Organic poultry processing on-farm should be properly learned and carefully carried out. Here are some guidelines:

- Develop your own HACCP (hazard analysis and critical control point) plan. For information on developing and implementing an HACCP plan, see www.fsis.usda.gov/factsheets.
- At minimum, use stainless steel equipment, clean, tested water, lots of chilling ice, and hygienic practices.
- Insist on gentle and humane handling of birds to the very end.
- Ensure integrity of the processing work, from slaughter to packaging to freezing.

You can find good information about on-farm processing uniquely suited to the small organic broiler producer. ATTRA (Appropriate Technology Transfer for Rural Areas) has a good publication called *Small-Scale Poultry*

Processing by Anne Fanatico. The American Pastured Poultry Producers Association offers numerous examples of how small producers are processing and selling their own product.

Composting Poultry By-Products

Aside from the benefit of quality control when processing on-farm, there is also the benefit of the by-products. Most organic growers would never view poultry processing by-products as waste. They are a rich source of nutrients that should not leave the farm or be wasted with improper disposal. Properly collected and composted poultry offal, blood, feathers, feet, and heads are an excellent nutrient source rich in nitrogen that should not be lost. (For a full discussion of composting, see *Compost, Vermicompost, and Compost Tea: Feeding the Soil on the Organic Farm* by Grace Gershuny, a companion NOFA Handbook).

Methods differ, but the compost mix in this case typically includes three basic components: poultry, a carbon source, and manure. Purdue University has an excellent publication called *Composting Poultry Carcasses* that details a design for large poultry operations needing to compost hundreds of birds at any one time.[1] This system can easily be adapted to a small poultry farm and resembles the techniques used by many Northeast organic producers who process on-farm. It is a fast-composting system, based on a series of composting bins where the ingredients are layered, mixed, turned, and composted. Here are some guidelines:

- Build roofed bins (at least two—see below) on a concrete slab.
- Bin size should be large enough to accommodate a day's production (cull birds, offal, etc.).
- For every pound of poultry one cubic foot of space is needed in the bin.
- Site should be well-drained, graded, and elevated so no water will enter the unit.
- The compost mixture is made by adding the correct quantity of birds, manure, and straw to the primary or first box. Here is the basic formula for poultry compost:

Material	Parts by weight
Poultry	1
Manure/litter	1½
Straw (wheat)	1/10
Water if needed	0–½

- The formula should provide a carbon-to-nitrogen ratio of 15–23:1 and a moisture content of 55–56 percent—critical for rapid and complete composting.
- Unless litter or manure is extremely dry, no added water should be needed.
- Straw portion can be hay, corn stover, dried grass, bean pods, shavings, sawdust, wood chips, any similar material normally used as litter, or finished compost (up to 50 percent substitution can be made).
- The composter should be loaded in the manner shown here.

Layering of Poultry Offal Compost Pile

Top	Manure cap
	Layer of straw
	6-inch layer of manure
	Single layer of poultry (6 inches away from edges of pile)
	Layer of straw
	6-inch layer of manure
	Double layer of straw
Bottom	Concrete pad

- Once the pile is capped, monitor the temperature (daily readings).
- Pile should reach 135°F–160°F within two to four days; temperatures in excess of 130°F are needed.
- Once the temperature drops to 120°F (fourteen days) compost is turned into the second bin and recapped.
- After another fourteen days at the proper temperature the compost can be stored or applied to fields as manure.
- If it fails to heat or smells bad it is usually due to the material being

too wet; moisture excludes oxygen, and the compost becomes anaerobic. To remedy, turn and add dry litter.

- Poultry compost is similar in nutrients to manure, but lower in nitrogen and higher in phosphorus and potassium.
- Proper heating destroys coliform, salmonella, Newcastle disease or IB (infectious bronchitis).

THE AUTHOR'S EXPERIENCE:
On-Farm Poultry Processing[2]

Chicken processing has been very successful at Kingbird Farm with the support of friends and family. My comments are prefaced by the fact that nearly every slaughter was assisted by friends and family who come for the lunch and apparently the fun of it. It defies logic to have such an excellent crew of volunteers every year willing (and skilled) to assist in changing a feathered bird into a clean broiler fit for the most particular customer.

That said, processing is often the biggest hurdle small producers face when contemplating pastured poultry. The brooding seems manageable; the pasturing seems within reason; even the marketing might be a snap, but doing the actual killing of potentially hundreds of birds can be daunting. Of all the steps, however, the processing is the one we want complete control of so that the birds are handled gently, killed respectfully, and cleaned with the utmost cleanliness and care. It's the only way I can stand behind my product 100 percent.

Our processing techniques and skills have been refined over the years. We began by tackling 50 birds in a day with the most rudimentary equipment. Most of our knowledge came from doing a few home-processed roosters for our own consumption. We quickly realized we needed an efficient, smooth-running operation with skilled help. We designed better equipment, refined our skills, and increased our production without sacrificing quality control. We still limit ourselves to 100 birds a day, catching birds in the dark, having breakfast at 7:00 a.m., and expecting to be finished with everything by lunchtime.

The Law

Under the laws of my state (New York), the licensing of slaughterhouses "shall not apply to any person who slaughters not more than two hundred fifty turkeys or an equivalent number of birds of all other species raised by him on his own farm during the calendar year for which an exemption is sought (four birds of other species shall be deemed the equivalent of one turkey), provided that such person does not engage in buying or selling poultry products other than those produced from poultry raised on his own farm." This exemption allows us to raise and process for retail sale 800 broiler chickens and 50 turkeys each year. It does not, however, allow us to do improper or unclean processing just because it is on-farm. All small poultry producers owe it to themselves, other producers, and especially their customers to process in a clean, safe, respectful manner. This does not require a lot of fancy equipment, expensive structures, or sanitizing chemicals. It does require common sense, simple cleanliness, and honesty.

Facility

We began, and have continued, with a very simple processing facility. We have seen many other small facilities over the years—garages, sheds, old milk houses, trailers and even a refitted school bus. We began with an open-sided 8 × 16 ft. shed with a gravel floor and have stayed there. The fresh air and sunshine make it a pleasure to work in (in most seasons) and offer the natural cleaning agent of weather. During foul weather—particularly likely during turkey processing—we can enclose the shed in tarps and even install a heater and lights. The setup allows for easy clean-up, minimal expense, and no place for pests to take up residence.

Equipment

Dependable, smooth-running processing equipment is essential. Our initial equipment was purchased secondhand out of old barns, but we quickly tired of its low capacity and tendency to break down. We keep the original scalder and plucker around as back-up, but have moved on to custom-built models.

On the kill side of the shed (divided by sheet metal walls and a curtain) my husband Michael has the most "complicated" equipment. There is a line of four metal "kill cones" on a sheet metal wall with blood buckets placed below. On the other side is a propane scalder with a rope and pulley for dunking and a barrel plucker. The scalder is simply a 50-gallon steel drum (food grade) with the top third cut off. It sits on a propane burner (turkey fryer) with a 5-gallon propane tank. Michael monitors the scald temperature (145°F) with a digital thermometer and adjusts the heat manually. The plucker was also handmade using a plastic drum and a rotating plate with fingers built within a wooden frame and powered by a small electric motor. This setup easily handles four birds at a time, approximately thirty an hour, a comfortable rate for Michael and the folks on the other side of the curtain.

On the clean side of the shed there are typically two of us eviscerating, cleaning, and bagging on an 8 ft. stainless steel table. I have done nearly all the eviscerating over the years since my small hands work well in the bird, and once someone is fast at eviscerating you don't want to replace him or her. I have simple equipment involving only 3 feet of tabletop with a hole for offal that drops into a bucket. On my counter is a tub of ice water for giblets, a carbon steel knife, a lung remover, and an overhead hose with a spray nozzle. The cleaning crew at the other end of the table has only pinning knives (for removing pinfeathers) and spray nozzles. When they are done cleaning, inspecting, and rinsing a bird it is transferred to the 100-gallon galvanized chill tank filled with ice water and covered by a clean tarp. For bagging we use a draining system of a series of rigid plastic pipes inserted vertically into a wooden stand. Birds are placed on the pipes to be drained of water and covered by a bag. We use freezer poultry bags, twist ties, and preprinted freezer labels that are filled out using a digital scale. On the clean side we also use plastic aprons, hats, and much hand washing. After handling, the birds are transferred in coolers to the freezers in our store at the barn.

The Process

The process begins the night before. Food is removed from the feeders, leaving the birds with only plenty of water and grass. We drive wooden poultry crates to the pasture for morning loading. We scrub down the chill tank with bleach and fill it half full of water (in order to save water the following day). We also pre-write the labels with price, date, and product to save time during bagging and weighing.

The next morning before light Michael and I load the birds into crates and drive them to the processing shed. He heats up his scald water and hooks up all the hoses (hot and cold water comes from the house). We try to start the processing by 8:00 a.m., both so the birds don't have to wait around and so we are finished by the time customers arrive. While Michael is killing the first four birds I scrub down the countertops, clean and sharpen knives, and prepare the ice water. I will usually see my first birds for eviscerating within five minutes.

On the kill side Michael keeps the crated birds out of the rain or hot sun. He selects from different crates, progressively giving all the remaining birds more room as he goes. Each bird is placed head-down in the kill cone. This usually calms them right down so Michael can quickly and cleanly cut their jugular. Bleed-out usually takes about four minutes; meanwhile he is scalding the previous batch. After the bleed-out, all four birds are secured by the feet to a pulley and dunked into the scalder (ninety seconds at 140°F) slowly several times. Michael uses dish soap in the water to break surface tension and penetrate the feathers. After scalding he turns on the plucker and places all four birds in at once for plucking, which takes approximately twenty seconds. He pulls any remaining large feathers and removes the head and feet. At this point the birds are transferred to the clean side of the shed.

Back on the clean side I begin the eviscerating process by removing the oil gland with a sharp knife. I then cut the neck skin and loosen the crop. I carefully open up the abdomen, remove the intestines, pull out the crop, and cut around the vent. All of this offal is dropped directly through a hole into the "gut bucket." I then remove the heart and liver and carefully pinch off the gall bladder. The heart and liver are kept in ice water until bagging. I use the lung remover to pull the lungs and a knife to remove the neck (which is also kept). After a good internal and external spraydown I pass the bird to the cleaning crew. They meticulously examine the bird, removing any pinfeathers, external fat, or blemishes. The birds are again rinsed inside and out before being placed in the chill tank. Our chill tank easily holds 50 gallons of cold water, 400 pounds of cube ice (4 lbs./bird) and 100 chickens. After chilling for at least an hour, the birds are drained and bagged. They are ferried up to the store in coolers where they are weighed, labeled, and frozen or placed in the commercial cooler for fresh pickup. During the entire process we strive to keep the chickens as cold (35°F–40°F) and clean as possible.

During cleanup, Michael's side is scrubbed and bleached. The buckets of feathers, heads, feet, blood, and offal are hauled up to the composting bins where they are layered with horse manure and straw (unturned, it makes a divine compost after two years). He covers his equipment and cleans up all the buckets. I scrub down and bleach all of the equipment on my side and drain the chill tank. The knives are cleaned and dried and the aprons are washed in the washing machine. I place everything on the counters to dry, air out, and get a little sunshine. Then we eat lunch.

Training

It can be overwhelming to start processing alone. I highly recommend getting as much experience as possible first: learn at the feet of the masters. In my case I was lucky to learn the finer points from Joel Salatin at his farm while processing 200 birds with his family. Books, articles, and chat rooms cannot replace getting your hands dirty next to an experienced farmer. Many times they may only let you observe, but that too will be very educational. Anything you do to increase your skills and respect for the process will benefit everyone in the end.

---

eight

Marketing Organic Poultry

If sound, humane, and healthy organic methods are used to grow poultry, the main purpose of marketing is to let the consumer know that. Consumers of organic poultry want to know that the birds were raised with care and compassion. They want to know that they received high-quality organic feed and had room to roam. They want the freshest product possible, and they want it to be delicious. Organic growers can offer all of these things plus innovation, dedication, and, above all, integrity. Small organic growers know that organic and free-range are not merely marketing ploys; they are farming methods with a deep integrity.

Packaging and Labeling

When packaging an organic product it is important to consider what materials the product will come in contact with and how it will be stored. Most FDA-approved plastic wraps or storage bags are made from three types of plastic: polyethylene (PE), polyvinlidene chloride (PVCD), or polyvinyl chloride (PVC). Plasticizers, colorants, or antifogging compounds may also be added. Even standard freezer paper is coated on one side with plastic to keep air out and prevent loss of moisture and freezer burn. PVC is usually preferred for wrapping meat as it protects against moisture loss, but has some oxygen permeability so allows meat to "bloom" (stay red and fresh looking). Shrink-wrapping involves a PVC film that shrinks when heated producing a tight fit. Vacuum packaging removes most of the air from containers before sealing in order to preserve flavor and retard bacterial growth. (*Note*: Some of these bags may be dusted with cornstarch for easy handling; this could easily be from GMO corn). The foam trays that meat is commonly wrapped on are made from expanded polystyrene (EPS). This material is made by adding foaming agents to polystyrene and passing it through a die. It is also used to make foam egg cartons.

With all these materials, small amounts of chemicals from packaging materials can migrate into foods. Some are not heat stable, and chemicals from plastic may migrate into food if left on during heating or thawing (particularly in a microwave).[1] No packaging is benign. Even white butcher/freezer paper has a plastic coating, so it's important to be informed about which packaging materials are appropriate for your intended use.

It is common among small egg producers to recycle egg cartons for resale. While this practice is quite noble in intent, it may be contaminating organic eggs. The Federal Safety Inspection Service (FSIS) considers egg cartons (pulp or foam) to be one-time-use packaging. Although it is typically legal for small backyard-flock owners selling directly to the public to reuse and relabel cartons, it is generally frowned upon in retail stores. Most stores purchasing eggs, particularly organic eggs, from small farmers will require the use of new, properly labeled cartons.

The very simplest labeling laws vary from state to state, but get more complicated when labeling USDA-inspected meats or eggs. It is important for a producer to research local and state law for labeling eggs or on-farm processed poultry. This is particularly important when selling off-farm at farmers' markets or when wholesaling to retail stores or restaurants.

Here are the New York Department of Agriculture and Markets labeling requirements for shell eggs (such requirements may vary by state):

1. The word "EGGS."
2. Grade—in letters ⅜ inch or larger.
3. Size—in letters ⅜ inch or larger.
4. Count—may be expressed in dozens.
5. Name, address, and zip code—if other than producer or packer must so state.
6. Identified as to source, with packer identification, or if labeled "Packed by," full name and address is acceptable.

Here are the New York labeling requirements for farm-processed poultry:

1. Product identification (broiler chicken, whole duck, etc.).
2. Slaughter date.

3. Name and address of producer.
4. EXEMPT—Art. 5A NYS DA&M Law.

And here are sample labeling requirements for USDA-inspected meat and poultry:

1. Product identification (i.e., Chicken).
2. Cut identification (i.e., Whole Fryer).
3. Plant number (USDA plant inspection number).

All of the minimum labeling requirements listed above involve simple, basic facts. Once a producer starts making claims on the label things get more complicated. The term "labeling" also includes any written, printed, or graphic material that is used on the containers or wrappings of poultry products, or that accompanies poultry products at the point of sale (i.e., "point of purchase" materials).[2]

"Organic" and Other Labeling Claims

It is permitted to label poultry with a factual statement that the product has been "certified organic by [name of certifying agency]." The term "organic" cannot be used by itself as a claim on the label except as part of the signature line on labels if the "organic" is part of the farm's incorporated name (e.g., Mike's Organic Poultry), and it is deemed not to be misleading. Any labeling statement that uses the term "organic" must be in accordance with the final rule and be submitted for prior approval to the Labeling and Additives Policy Division of the FSIS.

The FSIS has definitions for most labeling claims made by poultry producers on eggs and meat products. It is helpful to the consumer and to other producers if these terms are used honestly and consistently.

Rock Cornish Game Hen—small broiler 1–2 lbs.

Fryer—young birds less than 4 lbs.

Broiler—young birds 4–5 lbs.

Roaster—young birds 5–10 lbs.

Capon—male bird castrated at three weeks, 7–11 lbs.

Stewing Hen—older retired laying hen

Chemical Free—not allowed

No Antibiotics—may be used if producer can produce documentation proving this.

No Hormones—cannot be used unless it is followed by "Federal regulations prohibit the use of hormones."

Natural—a product containing no artificial ingredients or added color. The label must explain the use of the term.

Free-Range or Free-Roaming—producer must demonstrate to the agency (FSIS) that the poultry has been allowed access to the outside.

nine

Production Expectations and Economic Viability of Organic Poultry

It is important for a new producer to have some way to gauge success. Yes, the birds seem happy, but are they really producing well? Yes, the customers love the eggs, but are they really paying a fair price? Organic producers of all kinds chronically undervalue their products and fail to give themselves a living wage. Organic farming must be economically, as well as environmentally, sustainable. Farmers must value their time and their extraordinary efforts to provide humane and healthy poultry products. To meet financial expectations and charge a fair price, producers should study production benchmarks and cost-benefit analysis. Since production varies so dramatically by breed, location, rearing technique, and feed quality it is important to review production benchmarks from several sources. Carefully monitoring your own inputs and production numbers and comparing those with other producers who use similar methods may prove the most helpful.

Poultry Performance Benchmarks

Conventional layer operations work on a 25 dozen eggs/hen/year target. A reasonable target for free-range or organic producers would be 20 dozen/hen/year according to Maritime Certified Organic, a Canadian certifier.[1] Joel Salatin expects 21 dozen/hen/year with a drop of 25 percent in the winter.[2] Consult tables 11–15 for performance benchmarks on broilers, turkeys, ducks, and geese.

Budget Planning and Economic Viability

The success of any agricultural venture is based not only on the care of the animals and the quality of the product, but also the economic viability of the business. It is wise for farmers to complete a cost-benefit analysis as part of a marketing plan on any new or existing farm venture to determine its viability. (For planning help, see *Whole-Farm Planning* by Elizabeth Henderson and Karl North, and *The Organic Farmer's Guide to Marketing and Community Relations*, by Rebecca Bosch, both companion NOFA handbooks). This process includes many factors, from properly estimating overhead costs to charging a fair price. The Center for Integrated Agricultural Systems at the University of Wisconsin–Madison has developed a poultry-enterprise budget spreadsheet applicable to most kinds and sizes of poultry enterprises (see "Poultry Enterprise Budget," at www.wisc.edu/cias/pubs/poultbud.htm).[3] It takes the producer step-by-step through the definitions and structure of a poultry budget. Poultry scientists, producers, and other experts reviewed this budget system in

Table 11. Standards for Broiler Growth (Live Weight in Pounds)

Week	Cockerels	Pullets
1	0.27	0.25
2	0.61	0.54
3	1.09	0.94
4	1.66	1.41
5	2.24	1.89
6	2.91	2.42
7	3.66	3.00
8	4.38	3.58

Source: Based on information in Leonard S. Mercia, *Raising Poultry the Modern Way* (Charlotte, VT: Garden Way Publishing, 1975).

Table 12. Standards for Broad-Breasted White Turkey Growth– (Live Weight in Pounds)

Weeks	Toms	Hens
2	0.55	0 .53
4	1.40	1.20
6	2.50	2.00
8	4.00	3.50
10	5.40	5.20
12	8.20	7.00
14	11.00	9.00
16	14.00	11.00
18	16.90	13.00
20	19.70	14.80
22	22.40	16.30
24	25.20	17.30

Source: Based on information in Mercia, *Raising Poultry the Modern Way,* 112.

Table 13. Standards for White Pekin Ducklings–Mixed Sexes	
Weeks	Pounds liveweight
1	0.60
2	1.68
3	2.98
4	4.01
5	5.13
6	6.19
7	6.95
8	7.54

Source: Based on information in Mercia, *Raising Poultry the Modern Way,* 133.

Table 14. Growth Standards for Range-Reared Emden Goslings	
Weeks	Pounds liveweight
3	3.30
6	7.80
9	10.10
12	11.60
13	11.80

Source: Based on information in Mercia, *Raising Poultry the Modern Way,* 133.

Table 15. Poultry Processing Yields

Type	Approximate eviscerated weight as % of live
Broilers and Fryers	75
Roasters	76
Stew Hens (Leghorn Type)	68
Stew Hens (Heavy Type)	70
Turkeys (Heavy Roasters)	80
Turkeys (Light Roasters)	78
Ducks	70
Geese	68-73

Source: Based on information in Mercia, *Raising Poultry the Modern Way,* 161.

order to provide a thorough method for determining economic viability. Careful use of enterprise budgets such as this can provide data for better farm-business decision making.

Conclusion

Northeast organic poultry farmers have been and continue to be successful in the humane and healthy production of eggs and meat. Many of the producers surveyed and interviewed by this author have shown innovation and sincerity in their efforts to create integrated poultry systems that provide the birds with a natural life and themselves with a viable living. Most are small producers with flocks of fewer than a thousand who have included poultry in their farm's diversity. The job is more challenging for large organic producers, but they are also being creative in order to grow large numbers of birds under at least the minimum standards as currently defined. Both ends of the spectrum are adapting their systems always, one hopes, to reflect the spirit of the law and uphold the integrity of organic practices.

Appendices

Appendix 1: Housing and Equipment Sources

Ashley
901 N. Carver Street
PO Box 2
Greensburg, IN 47240
(812) 663-2180
Processing equipment

Brower Highway 16 West
PO Box 2000
Houghton, IA 52631
(318) 469-4141
www.browerequip.com
Feeders, drinkers, feed handling, poultry processing supplies

Carolina Fowlstuff
1381 Old Thomasville Road
Winston-Salem, NC 27107
(336) 769-4392
Fowlstuff@earthlink.net
Feeders, drinkers, processing supplies, housing supplies

EggCartons.com
24 Holt Road
Manchaug, MA 01526
(888) 852-5340
www.eggcartons.com
Egg cartons, boxes, feeders, drinkers, egg-washing supplies

Farmer Boy Ag Supply
410 East Lincoln Avenue
PO Box 435
Myerstown, PA 17067
(800) 845-3374
www.farmerboyag.com
Automatic feeding and watering systems, feeders, drinkers, brooders, feed-handling systems

FarmTek
14440 Field of Dreams Way
Dyersville, IA 52040
(800) 327-6835
www.farmtek.com
Housing structures, feeders, drinkers, brooders, automatic systems

Nasco Farm and Ranch
901 Janesville Avenue
Fort Atkinson, WI 53538
(800) 558-959
www.enasco.com
Feeders, drinkers, egg-handling supplies, processing supplies, brooders

Shenandoah Manufacturing Co.
PO Box 839, 1070 Virginia Avenue
Harrisonburg, VA 22801
(800) 476-7436
Gas brooders

Stromberg's
PO Box 400
Pine River, MN 56474
(800) 720-1134
www.strombergschickens.com
All things poultry; chicks, duck wax

Appendix 2: Fencing and Pasture Equipment Sources

Kencove
344 Kendall Road
Blairsville, PA 15717
(800) 536-2683
www.kencove.com
Chargers, polywire, testers

Lakeview Organic Grain
119 Hamilton Place
Penn Yan, NY 14527
(315) 531-1038
Certified organic seed

Premier
2031 300th Street
Washington, IA 52353
(800) 282-6631
www.premier1supplies.com
Poultry nets, fencing supplies, polywire, chargers

Wellscroft Fence Systems
167 Sunset Hill-Chesham
Harrisville, NH 03450
(603) 827-3464
www.wellscroft.com
Chargers, netting, polywire, supplies, technical support

Appendix 3: Hatcheries and Pullet Sources

Clearview Stock Farm and Hatchery
PO Box 399
Gratz, PA 17030
(717) 365-3234
Broilers, kosher kings, hybrid layers, standard layers, turkeys, ducks, geese, rare breeds, bantams, hatching eggs, equipment

Hoffman Hatchery Inc.
PO Box 129
Gratz, PA 17030
(717) 365-3694
www.hoffmanhatchery.com
Broilers, hybrid layers, standard layers, turkeys, ducks, geese, rare breeds, bantams, equipment

Johnson's Waterfowl
36882 160th Avenue NE
Middle River, MN 56737
(218) 222-3556
Ducks and geese

Kreamer Feed, Inc.
Kreamer, PA 17833
(800) 767-4537
www.organicfeed.com
Certified organic pullets

Marti Poultry Farm
PO Box 27
Windsor, MO 65360
(660) 647-3157
Martip@iland.net
Broilers, hybrid layers, standard layers, turkeys, ducks, geese, rare breeds, bantams

Mt. Healthy Hatcheries
9839 Winston Road
Mt. Healthy, OH 45231
(800) 451-5603
www.mthealthy.com
Broilers, hybrid layers, standard layers, turkeys, ducks, geese

Moyer's
266 E. Paletown Road
Quakertown, PA 1895
(215) 536-3155
www.moyerschicks.com
Broilers, hybrid layers, started pullets (potentially organic)

Privett Hatchery
PO Box 176
Portales, NM 88130
(505) 356-6425
Yucca.net/privetthatchery
Broilers, hybrid layers, standard layers, turkeys, ducks, geese, rare breeds, bantams

Reich Poultry Farms
1625 River Road
PO Box 100
Marietta, PA 17547
(717) 426-3411
Broilers, barred silver cockerels, hybrid layers, turkeys, ducks, geese

Townline Hatchery
PO Box 108
Zeeland, MI 49464
(616) 772-6514
Heritage turkey poults

Walters Hatchery
Rural Route 3, Box 1409
Stillwell, OK 74960
(918) 778-3535
www.historicalturkeys.com

Welp's, Inc.
PO Box 77
Bancroft, IA 50517
(800) 458-4473
www.welphatchery.com
Broilers, hybrid layers, standard layers, turkeys, ducks, geese, rare breeds, bantams

Appendix 4: Certified Organic Feed, Grains, Supplements, and Health-Care Products

Crystal Creek
N9466 Lakeside Road
Trego, WI 54888
(888) 376-6777
www.crystalcreeknatural.com
Probiotics herb and nutritional supports, technical support

Fertrell
PO Box 265
Bainbridge, PA 17502
(717) 367-1566
www.fertrell.com
Poultry Nutri-Balancer, fish and crab meal, probiotics, kelp, soil amendments, technical support

Green Mountain Feeds
PO Box 505
Bethel, VT 05032
(802) 234-6278
Certified organic poultry feeds and grains

Helfter Feeds, Inc.
136 N. Railroad Street,
PO Box 266
Osco, IL 61274
(866) 435-3837
www.helfterfeeds.com
Supplements and premixes

Kreamer Feed, Inc.
Kreamer, PA 17833
(800) 767-4537
www.organicfeed.com
Certified organic poultry feeds

Lakeview Organic Grain
119 Hamilton Place
Penn Yan, NY 14527
(315) 531-1038
Certified organic and transitional poultry feeds, Fertrell and Crystal Creek Dealer

McGeary Organics, Inc.
PO Box 299
Lancaster, PA 17608
(800) 624-3279
www.mcgearyorganics.com
Certified organic poultry feeds and grains

Northern-Most Feeds, LLC
155 Gagnon Road
Madawaska, ME 04756
(207) 728-3150
Northernmostfeeds@yahoo.com
Certified organic poultry feeds grown in Maine

Organic Unlimited
PO Box 238
120 Liberty Street
Atglen, PA 19310
(610) 593-2995
Certified organic poultry feeds

Appendix 5: Expert Consultants in Poultry

Jeff Mattocks, Fertrell, poultry nutrition. Jeff@fertrell.com

Robert Plamondon, Norton Creek Farm and Press, pastured poultry, egg handling, chick brooding. Robert@plamondon.com

Jo Robinson, author, *Why Grassfed Is Best*, grassfed poultry. Jor@teleport.com

Jim McLaughlin, Cornerstone Farm Ventures, pastured poultry. www.cornerstone-farm.com

Anne Fanatico, National Center for Appropriate Technology, free-range poultry. Annef@ncat.org

Paul Patterson, PSU Poultry Science, poultry science. Php1@psu.edu

Appendix 6: Sample Feed Recipes

Organic Layer Mash, No Corn (from Northern-Most Feeds)

Crude protein 16.5% min

Ingredients: Organic hull-less oats, Organic oats, Organic soy meal, Limestone, Organic vegetable oils, Organic vitamin and mineral supplements, Dicalcium phosphate, Salt, Spirulina

Organic Chick Starter Mash, No Corn (from Northern-Most Feeds)

Protein 20% min

Ingredients: Organic hull-less oats, Organic soy meal, Organic vegetable oils, Organic vitamin and mineral supplements, Dicalcium phosphate, Limestone, Salt, Spirulina

Pasture Layer Sample Ration—First Laying Cycle
(from Jeff Mattocks)

Crude protein 16.5% min

Ingredients:	*Lbs./ton:*
Alfalfa meal	100
Aragonite	175
Corn	965
Oats	100
Vitamin mineral premix	60
Roasted soybeans	600

Pasture Broiler Sample Ration (from Jeff Mattocks)

Crude protein 19.4% min

Ingredients:	*Lbs./ton:*
Alfalfa meal	100
Aragonite	25
Corn	1,015
Crab meal	75
Oats	100
Vitamin mineral premix	60
Roasted soybeans	2,000

Pasture Turkey Sample Starter Ration (from Jeff Mattocks)

Crude protein 25.7% min

Ingredients:	*Lbs./ton:*
Alfalfa meal	100
Aragonite	15
Corn	295
Dicalcium phosphate	10
Crab meal	100
Oats	250
Vitamin mineral premix	80
Roasted soybeans	950
Wheat	200

Certified Organic Layer Mash (from Kingbird Farm)

Crude protein 16% min

Ingredients:	*Lbs./ton:*
Corn	965
Wheat	200
Roasted soybeans	475
Calcium	150
Fertrell Nutribalancer	60
Crab meal	150

Certified Organic Broiler Mash (from Kingbird Farm)

Crude protein 18% min

Ingredients:	*Lbs./ton:*
Corn	1,010
Wheat	200
Roasted soybeans	630
Calcium	24
Fertrell Nutribalancer	60
Crab meal	76

Appendix 7: The National Organic Program Final Rule

§ 205.236 Origin of livestock

(a) Livestock products that are to be sold, labeled, or represented as organic must be from livestock under continuous organic management from the last third of gestation or hatching: Except, That,

(1) Poultry. Poultry or edible poultry products must be from poultry that has been under continuous organic management beginning no later than the second day of life;

(2) Dairy animals. Milk or milk products must be from animals that have been under continuous organic management beginning no later than 1 year prior to the production of the milk or milk products that are to be sold, labeled, or represented as organic, Except, That, when an entire, distinct herd is converted to organic production, the producer may:

(i) For the first 9 months of the year, provide a minimum of 80-percent feed that is either organic or raised from land included in the organic system plan and managed in compliance with organic crop requirements; and

(ii) provide feed in compliance with § 205.237 for the final 3 months

(iii) Once an entire, distinct herd has been converted to organic production, all dairy animals shall be under organic management from the last third of gestation.

(3) Breeder stock. Livestock used as breeder stock may be brought from a nonorganic operation onto an organic operation at any time: Provided,

That, if such livestock are gestating and the offspring are to be raised as organic livestock, the breeder stock must be brought onto the facility no later than the last third of gestation

(b) The following are prohibited:

(1) Livestock or edible livestock products that are removed from an organic operation and subsequently managed on a nonorganic operation may be not sold, labeled, or represented as organically produced.

(2) Breeder or dairy stock that has not been under continuous organic management since the last third of gestation may not be sold, labeled, or represented as organic slaughter stock

(c) The producer of an organic livestock operation must maintain records sufficient to preserve the identity of all organically managed animals and edible and nonedible animal products produced on the operation.

§ 205.237 Livestock feed

(a) The producer of an organic livestock operation must provide livestock with a total feed ration composed of agricultural products, including pasture and forage, that are organically produced and, if applicable, organically handled: Except, That, nonsynthetic substances and synthetic substances allowed under § 205.603 may be used as feed additives and supplements

(b) The producer of an organic operation must not:

(1) Use animal drugs, including hormones, to promote growth;

(2) Provide feed supplements or additives in amounts above those needed for adequate nutrition and health maintenance for the species at its specific stage of life;

(3) Feed plastic pellets for roughage;

(4) Feed formulas containing urea or manure;

(5) Feed mammalian or poultry slaughter by-products to mammals or poultry; or

(6) Use feed, feed additives and feed supplements in violation of the Federal Food, Drug and Cosmetic Act

§ 205.238 Livestock health-care practice standard

(a) The producer must establish and maintain preventive livestock health-care practices, including:

(1) Selection of species and types of livestock with regard to suitability for site-specific conditions and resistance to prevalent diseases and parasites;

(2) Provision of a feed ration sufficient to meet nutritional requirements, including vitamins, minerals, protein and/or amino acids, fatty acids, energy sources and fiber (ruminants);

(3) Establishment of appropriate housing, pasture conditions and sanitation practices to minimize the occurrence and spread of diseases and parasites;

(4) Provision of conditions which allow for exercise, freedom of movement, and reduction of stress appropriate to the species;

(5) Performance of physical alterations as needed to promote the animal's welfare and in a manner that minimizes pain and stress; and

(6) Administration of vaccines and other veterinary biologics

(b) When preventive practices and veterinary biologics are inadequate to prevent sickness, a producer may administer synthetic medications: Provided, That, such medications are allowed under § 205.603. Parasiticides allowed under § 205.603 may be used on

(1) Breeder stock, when used prior to the last third of gestation but not during lactation for progeny that are to be sold, labeled, or represented as organically produced; and

(2) Dairy stock, when used a minimum of 90 days prior to the production of milk or milk products that are to be sold, labeled, or represented as organic

(c) The producer of an organic livestock operation must not:

(1) Sell, label, or represent as organic any animal or edible product derived from any animal treated with antibiotics, any substance that contains a synthetic substance not allowed under § 205.603, or any substance that contains a nonsynthetic substance prohibited in § 205.604

(2) Administer any animal drug, other than vaccinations, in the absence of illness;

(3) Administer hormones for growth promotion;

(4) Administer synthetic parasiticides on a routine basis;

(5) Administer synthetic parasiticides to slaughter stock;

(6) Administer animal drugs in violation of the Federal Food, Drug and Cosmetic Act; or

(7) Withhold medical treatment from a sick animal in an effort to preserve its organic status. All appropriate medications must be used to restore an animal to health when methods acceptable to organic production fail. Livestock treated with a prohibited substance must be clearly identified and shall not be sold, labeled, or represented as organically produced.

§ 205.239 Livestock living conditions

(a) The producer of an organic livestock operation must establish and maintain livestock living conditions which accommodate the health and natural behavior of animals, including:

(1) Access to the outdoors, shade, shelter, exercise areas, fresh air and direct sunlight suitable to the species, its stage of production, the climate and the environment;

(2) Access to pasture for ruminants;

(3) Appropriate clean, dry bedding. If the bedding is typically consumed by the animal species, it must comply with the feed requirements of § 205.237;

(4) Shelter designed to allow for:

(i) Natural maintenance, comfort behaviors and opportunity to exercise;

(ii) Temperature level, ventilation and air circulation suitable to the species; and

(iii) Reduction of potential for livestock injury;

(b) The producer of an organic livestock operation may provide temporary confinement for an animal because of:

(1) Inclement weather;

(2) The animal's stage of production;

(3) Conditions under which the health, safety, or well being of the animal could be jeopardized; or

(4) Risk to soil or water quality

(c) The producer of an organic livestock operation must manage manure in a manner that does not contribute to contamination of crops, soil, or water by plant nutrients, heavy metals, or pathogenic organisms and optimizes recycling of nutrients.

Appendix 8: Recommended Web Sites

www.acornorganic.org
Maritime Certified Organic
Organic poultry management

www.fsis.usda.gov/oa/background/organic
Food Safety and Inspection Service
Organic labeling

www.extension.umn.edu
University of Minnesota Extension
Poultry processing

www.misa.umn.edu
University of Minnesota Extension
Organic poultry feed

www.poultryhelp.com
Links and articles about poultry, many organic

www.lionsgrip.com
Lions Grip company
Pastured poultry feed information

www.ansci.umn.edu
Animal Science Department, University of Minnesota
Organic poultry feedstuffs studies by Jaquie Jacobs

www.newfarm.org
Rodale Institute
Clearinghouse for alternative farming

www.localharvest.org
Directory of small farms
Good background information on products

www.feathersite.com
Poultry clearinghouse

www.vethomeopath.com
Dr. Surjit S. Makker
Homeopathy for poultry

www.veeru.reading.ac.uk
University of Reading
Poultry diagnostic and treatment manual

www.bfrepa.co.uk
British Free-range Egg Producers Association
Excellent articles and technical papers

www.eatwild.com
Jo Robinson
Grass fed and organic poultry listings

www.free-rangepoultry.com
Herman Beck-Chenoweth
Poultry articles

www.poultryconnection.com
Poultry clearinghouse

www.plamondon.com
Robert Plamondon
Brooding and management of pastured poultry

www.apppa.org
APPPA: The American Pastured Poultry Producers Association
Excellent articles and other information about grass-fed poultry

www.attra.ncat.org
ATTRA: Appropriate Technology Transfer for Rural Areas
Excellent publications

www.agmrc.org
Agricultural Marketing Resource Center
Marketing resource

Endnotes

Chapter 1

1. "Five Freedoms," Compendium of Animal Health and Welfare in Organic Farming, Organic Livestock Research Group, VEERU, Department of Agriculture, The University of Reading. 2000. www.organic-vet.reading.ac.uk. Also www.ruleworks.co.uk/poultry/five-freedoms.asp and the Animal Rights Encyclopedia: www.zoosavvy.com/five-freedoms.html.
2. "The Natural Growth of Organic Broilers," *Poultry* (Dec/Jan 2003), 24.

Chapter 2

1. Robert Plamondon, "Range Poultry Housing," ATTRA, 1999; www.attra.org.
2. Karma E. Glos, *Remedies for Common Health Problems of the Organic Laying Flock*, SARE Project FNE02-412, survey of Northeast organic poultry producers.
3. NOSB Livestock Committee, draft recommendation, "Access to the Outdoors for Poultry," December 21, 2001; www.ams.usda.gov/nosb/NOSBrecommendations/poultry.html.
4. Anne Macey, *Organic Livestock Handbook* (Ottawa: Mothersill Printing, 2000).
5. National Organic Standards Board, Compost Task Force, Final Report.
6. See USDA National Organic Program Regulation, Final Rule, section 205.203, subpart C (National Organic Program, 2000).
7. Originally published in *Organic Farms, Folks, and Foods*, The Northeast Organic Farming Association of New York (Midwinter 2003).
8. On the Country Hen controversy see *NOFA/Mass News* (Dec.–Jan. 2002–3), and *The Natural Farmer* (Winter 2002–3);
9. "Organic Still Means Humane," The Humane Society of the United States, www.hsus.org/ace/14057.
10. "HSUS Recommendations to the NOSB, Draft Recommendations of the NOSB Livestock Committee: Access to the Outdoors," The Humane Society of the United States, 2002, www.hsus.org/ace/14086.
11. "Organic Still Means Humane," The Humane Society of the United States, www.hsus.org/ace/14057.
12. Jo Robinson, *Why Grassfed Is Best* (Vashon, WA: Vashon Island Press, 2000), 33.

13. Heather Karsten and Paul Patterson, "Pasture-ized Poultry," http://www.rps.psu.edu/0305/poultry.html.
14. Ibid.
15. Ibid.
16. Ibid.
17. Herman Beck-Chenoweth, *Free-Range Poultry Production and Marketing*, p.3.1.

Chapter 3

1. Anne Fanatico and Skip Polson, "Which Bird Shall I Raise?: Genetic Options for Pastured Poultry Producers: Meat-type Chickens and Turkeys" (December 2002), www.apppa.org/apppa/resources.htm.
2. Donald Bixby, "The Importance of Genetics: Biological Fitness and Productivity in Range-Based Systems Comparing Standard Turkey Varieties and Industrial Stocks," SARE Project # LS02-134.
3. Originally published in *Organic Farms, Folks, and Foods*, The Northeast Organic Farming Association of New York (Mid-Spring 2002).

Chapter 4

1. Daniel Ungier, "Finding Options for Organic Poultry Feed," Minnesota Institute for Sustainable Agriculture (December 2002), www.misa.umn.edu/other/poultryfeed.html.
2. "Organic Feed for Poultry and Livestock: Availability and Prices," www.ams.usda.gov/nop/prodhandlers, 20.
3. Jeff Mattocks, *Pastured-Raised Poultry Nutrition*, Prepared for Heifer International, (SARE Project, 2002).
4. Ibid., appendix A5.
5. Ibid., 10.
6. www.poultrynotes.txt.

Chapter 5

1. Taylor Hyde, "How to Wean Yourself from the Vet," *Holistic Management Quarterly* (Summer 1997).
2. Anne Fanatico, "Sustainable Chicken Production," ATTRA, www.attra.org, 8.
3. Damerow, *Chicken Health Handbook*, 286.
4. Macey, *Organic Livestock Handbook*, 15.
5. Burcombe Organic Hatchery, www.organic-chicks-poultry.com.
6. Originally published in *Organic Farms, Folks, and Food*, The Northeast Organic Farming Association of New York (Mid-Spring 2004).

Chapter 6

1. Mattocks, *Pasture-Raised Poultry Nutrition*, 10.
2. Originally published in *Organic Farms, Folks, and Foods*, The Northeast Organic Farming Association of New York (Mid-Winter 2002).
3. Originally published in *Organic Farms, Folks, and Foods*, The Northeast Organic Farming Association of New York (Mid-Summer 2003).
4. Anne Fanatico, "Small-Scale Poultry Processing," ATTRA, 2003; www.attra.ncat.org.
5. Robert Plamondon, "Egg-Washing Procedure," www.poultrynotes.txt.
6. www.bfrepa.co.uk/pale_eggs.htm (April 2002).
7. www.bfrepa.co.uk.
8. Jacqueline Jacobs and Richard Miles, "Designer and Specialty Eggs" (University of Florida Extension, Fact Sheet PS-51, 2002).
9. "What's in an Egg (and What Shouldn't Be)," www.bfrepa.co.uk/reference.htm (January 2000).
10. Corby Kummer, "A Better Egg," *The Atlantic Monthly* (October 2000), 120.
11. Leslie Zack, "Pennsylvania Is Home to a Lot of Turkeys—Organic Turkeys, That Is," *Organic Matters*, Pennsylvania Certified Organic Quarterly Newsletter 7, no. 3 (Autumn 2003), 1.
12. Originally published in *Organic Farms, Folks, and Foods*, The Northeast Organic Farming Association of New York (Mid-Fall 2002).

Chapter 7

1. Dick Adams, Cal Flegal, and Sally Noll, *Composting Poultry Carcasses* (Purdue University, NCR-530, 1994).
2. Originally published in *Organic Farms, Folks, and Foods*, The Northeast Organic Farming Association of New York (Mid-Summer 2002).

Chapter 8

1. "Meat Packaging Materials," Food Safety and Inspection Services (FSIS), www.fsis.usda.gov/oa/pubs/meatpack.htm (May 2000).
2. "Using the Claim 'Certified Organic By . . .' On Meat and Poultry Product Labeling," FSIS, www.fsis.usda.gov/oa/background/organic.htm (March 2000).

Chapter 9

1. Roger Henry, "Organic Poultry-Eggs," Maritime Certified Organic Growers, "Organic Profiles" (March 2002).
2. Anne Fanatico, "Sustainable Poultry: Production Overview," ATTRA, www.attra.ncat.org/attra-pub/poultryoverview.html, 2.
3. "Poultry Enterprise Budget," Center for Integrated Agricultural Systems, www.wisc.edu/cias/pubs/poultbud.htm.

Bibliography

Adams, Dick, Cal Flegal, and Sally Noll. *Composting Poultry Carcasses.* Purdue University, NCR-530, 1994.

Agricultural Alternatives. *Small-Scale Egg Production (Organic and Nonorganic).* College of Agricultural Sciences, Pennsylvania State University, Cooperative Extension, 1999.

Agricultural Marketing Service. *Organic Feed for Poultry and Livestock: Availability and Prices.* USDA, 2003. www.ams.usda.gov/nop/prod handlers.

American Livestock Breeds Conservancy. albc@albc-usa.org.

Bairacli Levy, Juliette. *The Complete Herbal Handbook for Farm and Stable.* London and Boston: Faber and Faber, 1952.

Beck-Chenoweth, Herman. *Free-Range Poultry Production & Marketing.* Creola, OH: Back Forty Books, 1996.

Bixby, Donald. *The Importance of Genetics: Biological Fitness and Productivity in Range-based Systems Comparing Standard Turkey Varieties and Industrial Stocks*, SARE Project #LS02-134, 2002.

British Free-range Egg Producers Association 2002. www.bfrepa.co.uk.

Certified Organic Associations of British Columbia. *British Columbia Certified Organic Production Operation Policies and Management Standards*, version 5, book 2, (March 2003).

Damerow, Gail. *The Chicken Health Handbook.* North Adams, MA: Storey Publishing, 1994.

Dennett, Mick. *Profitable Free Range Egg Production.* Wiltshire, UK: The Crowood Press, 1995.

El-Begearmi, Mahmoud, *Amounts to Feed Your Flock.* Poultry Facts, Bulletin #2051, University of Maine Cooperative Extension, 2001.

Fanatico, Anne. "Feeding Chickens." ATTRA, 1998. www.attra.ncat.org/ attra-pub/feeding.html.

———. "Sustainable Chicken Production." ATTRA, 1998. www.attra.org.

———. "Range Poultry Housing." ATTRA,1999. www.attra.org.

———. "Sustainable Poultry: Production Overview." ATTRA, 2002. www.attra.ncat.org/attra-pub/poultryoverview.html.

———. "Small-Scale Poultry Processing." ATTRA, 2003. www.attra.ncat .org.

———. "Egg Processing by Hand." APPPA Grit, Issue 26.

Fanatico, Anne, and Skip Polson. "Which Bird Shall I Raise: Genetic Options for Pastured Poultry Producers: Meat-type Chickens and Turkeys." APPPA, 2002. www.apppa.org/apppa/resources.htm.

Food Safety and Inspection Service. "Meat Packaging Materials." United States Department of Agriculture, 2000. www.fsis.usda.gov.

———. "Using the Claim 'Certified Organic By . . .' on Meat and Poultry Product Labeling." United States Department of Agriculture, 2000. www.fsis.usda.gov.

———. "Meat and Poultry Labeling Terms," United States Department of Agriculture, 2003. www.fsis.usda.gov.

———. "Focus On: Chicken," United States Department of Agriculture, 2003. www.fsis.usda.gov.

Franczyk, Don. *Research on Pastured Poultry Breeds—Summer 2002 Special Supplement on On-Farm Research. The Natural Farmer*, Northeast Organic Farming Association (Summer 2002).

Glos, Karma E., ed. *Remedies for Health Problems of the Organic Laying Flock: A Compendium and Workbook of Management, Nutritional, Herbal, and Homeopathic Remedies*. SARE Project FNE02-412, 2003.

Grow, Oscar. *Modern Waterfowl Management and Breeding Guide*. American Bantam Association, 1972.

Hamre, Melvin L. *Evaluating Egg Production Hens*. University of Minnesota Extension Service, FS-1182-GO, 1990.

Holderread, Dave. *Raising the Home Duck Flock*. Charlotte, VT: Garden Way Publishing, 1978.

Humane Society of the United States. 2003. www.hsus.org.

Hyde, Taylor. "How to Wean Yourself from the Vet." *Holistic Management Quarterly* (Summer 1997).

Jacobs, Jacqueline, and Richard Miles. "Designer and Specialty Eggs." University of Florida Extension, Fact Sheet PS-51, 2002.

Karsten, Heather, and Paul Patterson. "Effect of Legume and Grass Pasture Species on Omega Fat, Vitamin A & E Concentration in Hen Eggs." Crop and Soil Sciences and Poultry Science Departments, The Pennsylvania State University, http://cropsoil.psu.edu/people/faculty/Karsten/PastPoultryPASA_files.

Kummer, Corby. "A Better Egg." *Atlantic Monthly* (October 2000): 118–22.

Lacey, Michael P. "Litter Quality and Broiler Performance." The University of Georgia College of Agricultural and Environmental Sciences, 2003. www.ces.uga.edu/pubcd/L426-w.html.

Lee, Clarence E. *Profitable Poultry Management*. Cayuga, NY: The Beacon Milling Company, Inc., 1928.

Luening, Robert A., and Donald J. Schuster. "Poultry Enterprise Budget." Center for Integrated Agricultural Systems, 2004. www.wisc.edu/cias/pubs/poultbud.

Macey, Anne, ed. *Organic Livestock Handbook: A Project of Canadian Organic Growers Inc.* Ottawa: Mothersill Printing, 2000.

Makker, Surjit S. "Homeopathic Treatment of Animals: Poultry Diseases." 2003. vethomeopath.com/poultry.

Maritime Certified Organic Growers. "Organic Profiles," 2002. www.acornorganic.org.

Mattocks, Jeff. *Pastured-Raised Poultry Nutrition.* Prepared for Heifer International, SARE Project, 2002.

Mercia, Leonard S. *Raising Poultry the Modern Way.* Charlotte, VT: Garden Way Publishing, 1975.

National Organic Standards Board. www.ams.usda.gov/nosb.

Organic Livestock Research Group. *Compendium of Animal Health and Welfare in Organic Farming.* VEERU, Department of Agriculture, University of Reading, 2000. www.organic-vet.reading.ac.uk.

Pennsylvania State University, College of Agricultural Sciences, Cooperative Extension. "Small-Scale Egg Production (Organic and Nonorganic)." 1999. http://agalternatives.aers.psu.edu/Publications/small_scale_egg.pdf.

Plamondon, Robert. "Mixing Free-Range Hens and Ruminants on Pasture." Norton Creek Farm, Slide Presentation, 2000.

———. "Egg-washing Procedure." Poultrynotes.txt chat room via Robert@plamondon.com, 2002.

———. *Success with Baby Chicks.* Blodgett, OR: Norton Creek Press, 2003.

Powell, Edwin C. *Making Poultry Pay.* New York: Orange Judd Company, 1904.

Roberts, Michael. *Modern Free Range.* The Gold Cockerel Series. Worcester, UK: The Roman Press, 1988.

Roberts, Victoria. *Disease of Free-range Poultry.* Suffolk, UK: Whittet Books Ltd., 2000.

Robinson, Jo. *Why Grassfed Is Best.* Vashon, WA: Vashon Island Press, 2000.

Salatin, Joel. *Pastured Poultry Profits.* Swoope, VA: Polyface, Inc., 1993.

Shane, Simon. "The 'Natural' Growth of Organic Broilers." *Poultry* (December/January 2003).

Singh, Av. "Managing Internal Parasites in Organic Livestock." Organic Agriculture Centre of Canada, 2003. asingh@nsac.ns.ca.

Smith, Miranda, and Elizabeth Henderson, eds. 1998. *The Real Dirt: Farmers Tell About Organic and Low-Input Practices in the Northeast.* Burlington, VT: The Northeast Region SARE/ACE Programs, 1998.

Soil Association. *Soil Association Question and Answers.* Bristol, UK: Soil Association Producer Services, 2003.

Thear, Katie. *Free-range Poultry.* Ipswich, UK: Farming Press, 1999.

Thompson, Alan. *Poultry Farming.* New York: Penguin Books, 1945.

Ungier, Daniel. "Finding Options for Organic Poultry Feed." Minnesota Institute for Sustainable Agriculture (December 2002). www.misa.umn .edu/other/poultryfeed.html.

VEERU, Organic Livestock Research Group, Department of Agriculture, University of Reading, 2000. www.organic-vet.reading.ac.uk.

Zack, Leslie. "Pennsylvania Is Home to a Lot of Turkeys—Organic Turkeys, That Is." *Organic Matters*, Pennsylvania Certified Organic Quarterly Newsletter 7, no. 3 (Autumn 2003), 1.

Index

About the Author and Illustrator

Karma Glos is the daughter of homesteading parents. She majored in field biology at Evergreen State College in her home state of Washington and worked briefly for the U.S. Forest Service studying small mammal ecology. With her husband, Michael, she has been building Kingbird Farm in upstate New York since 1996. Their successful, diverse operation includes hogs, poultry, beef, vegetables, and herbs with the help of many draft horses. She owns she has "a passion for healthful, innovative livestock husbandry practices."

Illustrator Jocelyn Langer is an artist, music teacher, and organic gardener, and the illustrator of the NOFA organic-farming handbooks in this series. She illustrates and does graphic design work for alternative media and political events as well as organic-farming-related publications. Jocelyn lives in central Massachusetts.

Julie Rawson was the special farmer-reviewer for this manual, and Michael J. Ghia was the scientific reviewer.